Art Meets Mathematics in the Fourth Dimension

Stephen Leon Lipscomb

Art Meets Mathematics in the Fourth Dimension

Second Edition

Stephen Leon Lipscomb
Emeritus Professor of Mathematics
University of Mary Washington
Fredericksburg, VA, USA

Gerald Edgar suggested the *Art Meets Mathematics in the Fourth Dimension* title.

Videos to this book can be accessed at
http://www.springerimages.com/videos/978-3-319-06253-2

ISBN 978-3-319-06253-2   ISBN 978-3-319-06254-9 (eBook)
DOI 10.1007/978-3-319-06254-9
Springer Cham Heidelberg New York Dordrecht London

Library of Congress Control Number: 2014940943

Mathematics Subject Classification: 28A80, 55M10, 97M80

Springer International Publishing Switzerland 2011, 2014
This work is subject to copyright. All rights are reserved by the Publisher, whether the whole or part of the material is concerned, specifically the rights of translation, reprinting, reuse of illustrations, recitation, broadcasting, reproduction on microfilms or in any other physical way, and transmission or information storage and retrieval, electronic adaptation, computer software, or by similar or dissimilar methodology now known or hereafter developed. Exempted from this legal reservation are brief excerpts in connection with reviews or scholarly analysis or material supplied specifically for the purpose of being entered and executed on a computer system, for exclusive use by the purchaser of the work. Duplication of this publication or parts thereof is permitted only under the provisions of the Copyright Law of the Publisher's location, in its current version, and permission for use must always be obtained from Springer. Permissions for use may be obtained through RightsLink at the Copyright Clearance Center. Violations are liable to prosecution under the respective Copyright Law.
The use of general descriptive names, registered names, trademarks, service marks, etc. in this publication does not imply, even in the absence of a specific statement, that such names are exempt from the relevant protective laws and regulations and therefore free for general use.
While the advice and information in this book are believed to be true and accurate at the date of publication, neither the authors nor the editors nor the publisher can accept any legal responsibility for any errors or omissions that may be made. The publisher makes no warranty, express or implied, with respect to the material contained herein.

Printed on acid-free paper

Springer is part of Springer Science+Business Media (www.springer.com)

Dedicated to

James Cornelius Perry
August 9, 1935 – January 10, 2011

Circa 2000, Perry created an isotopy that moves the 4-web from the 4th dimension into human view while preserving its fractal dimension and exposing its self-similarity.

AND

Eugene LeRoy Miller
November 5, 1939 – September 4, 2012

August 2000, Miller creates first physical-art representation of the 4-web.

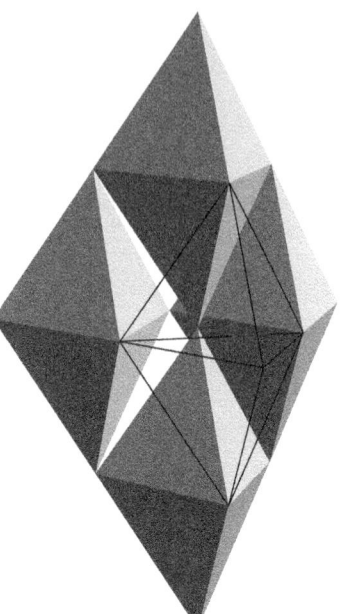

Basic 4-web Structure from the
Fourth-Dimension (Graphic by Chris Dupilka)

"The most beautiful thing we can experience is the mysterious.

It is the source of all true art and science."

Albert Einstein

As an instance of Einstein's statement, consider the mysterious question.

How can we use art to picture the intersection of a line and a 3-sphere?

Knowing that the 3-sphere is a 3-manifold, one may incorrectly guess

For a local view, picture a needle in an ice cube.

But (section) 40 shows that a line intersects a 3-sphere in at most two points. So the correct answer is

Three works of art (pictures) are required: one containing nothing, another containing one dot of ink, and another that contains two dots of ink.

The point is, pictures can convey "mysterious mathematical facts" to even non-mathematicians, and this truism is the keystone of this book.

Preface (Second Edition)

The title of the 2011 first edition is "Incipit" (here begins), and its subtitle is the title of this 2014 second edition. Other adjustments include the footnote on page 68 where "skull" replaces "scull"; the top of page 79 where "$-t(2\mathbf{m}\cdot\mathbf{b})$" replaces "$+t(2\mathbf{m}\cdot\mathbf{b})$"; the third paragraph on page 80; the middle of page 82 where the Pov-ray code-definition of "sphere" is corrected; and finally, the top-right of page 118 where "$(12/d)^2$" replaces "$(12/d)$".

The first edition contains a Blu-ray HD Disc, and MPG-4 versions of the videos therein are available at Springer web sites: The author's narration explains the coloring of the *God's Image?* art, with background music the art is rotated, and a 4-web grid from the 4th dimension is moved into human view. Frames of these videos are illustrated within the color-plate section.

For the creation of this second edition, the author gives thanks to Springer editors Ann Kostant and Marc Strauss for their guidance and support.

Spotsylvania, VA *Stephen L. Lipscomb*
April 2014

About the Author

A graduate of Fairmont State University BA; West Virginia University MA, and the University of Virginia Ph.D., where the late G. T. Whyburn (past president of the AMS) was his advisor. And prior to Chairing the Department of Mathematics, Physics, and Computer Science at the University of Mary Washington, Lipscomb was a U. S. Navy senior mathematician involved with the Trident Missile system and served as Team Chair of an Operational Evaluation of the Tomahawk Missile system.

In the 1970s the author solved a half-century old embedding problem in dimension theory. His solution, upon review by a Transactions of the American Mathematical Society (AMS) referee, was called "an outstanding contribution to dimension theory". In 2009 the author's feature article *The Quest for Universal Spaces in Dimension Theory* appeared in the December issue of the *Notices* of the AMS, and his book *Fractals and Universal Spaces in Dimension Theory* was published in the outstanding *Springer Monographs in Mathematics Series*. It turned out that the classical fractals known as Sierpiński's triangle and cheese were corollary to the author's solution. And at the beginning of the 21st Century, James Perry introduced another example (*the 4-web*) of the author's construction that lived in the Fourth Dimension. Perry's example is fundamental to this book.

During the 1990s, the author introduced an extension of Cauchy's *cycle notation* in group theory to *path notation* in semi-group theory. He used his path notation to construct new semi-groups, e.g., the *alternating semi-groups*. By 1996, the author's work with his path notation was documented by the AMS in Volume 46 *Symmetric Inverse Semigroups* of its prestigious *Mathematical Surveys and Monographs Series*.

Circa 2000 the author realized, because the triangle is the Holy Grail of strength design, that the 4-web would be ideal for practical applications. By 2005 he obtained a US patent and designed medical 4-web *spine cages*. Then Jessee Hunt created 4-Web, Inc., the cages eventually became FDA approved for human implantation, and the implants began in 2011. In 2014 the US Patent Office approved his 5-Web Patent.

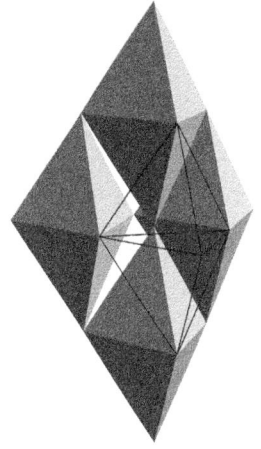 

Preface

Over time the impossible becomes understandable and partially possible.

Here begins *new art* created by *capturing pictures of objects that live in the 4th dimension.* The new method is applied to the *hyper-sphere*, which is a higher-dimensional offspring of an ordinary sphere. We see an ordinary sphere, also called a 2-*sphere*, each time we look at a full moon or the surface of a basketball, but the hyper-sphere lives in the 4th-dimension beyond human view. A hyper-sphere is also called a 3-*sphere*.

The "global structure" of a hyper-sphere requires four dimensions, but human vision is only 3-dimensional. In other words, it is humanly impossible to "see" a hyper-sphere in the 4th dimension.

Nevertheless, the historical value of the hyper-sphere cannot be overstated: Einstein used it in 1917 to describe the shape of our universe, and Dante, circa 1300 AD, used it to describe the Christian afterlife. These descriptions are presented in chapters 2, 3, and 4 using numerous illustrations.

Over time, our understanding of the meaning of *dimension* evolved. By 1970 a 4-*web* (think of a spider's web) from the 4th dimension emerged. In the 4-web case, however, by 2003 it was discovered that its "global structure" could be moved into human view (see left-side graphic above and reference [27]). The 4-web is an example of a construction that was applied in the article "On Imbedding Finite-Dimensional Metric Spaces", which I published in the *Transactions of the American Mathematical Society* Vol. 211, 1975.

While the 'human view" of the 4-web is a relatively recent discovery, the idea of "reasoning well with images of objects" dates back to the late 1800s. For example, within the introduction of Poincaré's famous *Analysis Situs* (1895), we find

> ... geometry is the art of reasoning well with badly made figures. Yes, without doubt, but with one condition. The proportions of the figures might be grossly altered but their elements must not be interchanged and must conserve their relative situation.

By moving the 4-web into human view, we "see" its structure: The line segments that comprise the 4-web are organized into *five groups*, each group "just touching" the other four. As Poincaré might say, the five groups were not interchanged and their relative situation was conserved.

Moving objects from one dimension into another begs the question *What do we mean when we say that space has three dimensions*? Such a question concerned Poincaré in 1912, when he stated

> ... Of all the theorems of analysis situs, the most important is that which we express by saying that space has three dimensions. ...

In this book the concept of *dimension* is tailored for a general audience. To accommodate such an audience, we introduce our friend *Freddy the penguin* whose dimensionally-varying vision is illustrated in numerous situations — Freddy motivates an intuitive understanding of dimension.

The *dimension* concept and the 4-*web* structure are fundamental to the generation of our hyper-sphere art. Roughly speaking, the graphic is generated in two steps: First, the 4-web is used to capture tens-of-thousands of points of a hyper-sphere, like a spider uses his web to capture insects. Second, the captured points are moved into human view using the algorithm for moving the 4-web into human view. The image points that define the graphic, being part of the 4-web graphic, meet Poincaré's requirement — they are *not interchanged and their relative situation is conserved.*

So given the modern, 21-Century mathematical, understanding of the 4-web and how it may be moved into human view, it is possible to create a *partial image of the hyper-sphere* (right-side graphic above) — *the impossible becomes partially possible.*

But what would a new work of art be without a title? In this case, the title reflects the history of the hyper-sphere and viewer impressions. Most viewers see the *face of man*, and such impressions combined with Einstein's and Dante's use of the hyper-sphere were key to the selection of the title "*God's Image?*" (with a question mark).

During the writing and rewriting of drafts of this book several people provided valued contributions: Some were colleagues, some are 3-D graphic specialists, and others were students. My former student Chris Edward Dupilka provided computer algorithms, beautiful graphics, and the first movie on the supplemental Blu-ray Disc; my colleagues Allen Danforth Parks, John Earl Gray, Eugene LeRoy Miller, Jimmy Gardner, Larry Lehman, and, Janet and Richard Zeleznock carefully read drafts and provided suggestions that greatly improved the presentation for a general audience; and the Rhino 3-D software (www.rhino3d.com) specialists Mary Fugier, Jerry Hambly, and Pascal Golay provided Rhino script and macros that supported the development of graphics and videos.

Spotsylvania, VA   *Stephen L. Lipscomb*
June 2011

Contents

Basic Structure of 4-web (1 illustration) vi

Preface (Second Edition) vii

Preface ix

Introduction (2 Illustrations) xv

Chapter 1. 3-Sphere (13 illustrations) 1
    §1 Spheres
    §2 Why is it that we cannot see a 3-sphere?
    §3 Spheres and discs
    §4 Gluing discs
    §5 Slicing spheres
    §6 Comments

Chapter 2. Dante's 3-Sphere Universe (10 illustrations) 11
    §7 Aristotle universe
    §8 Dante's journey
    §9 Angelic sphere
    §10 Dante's 3-sphere
    §11 Locally glued areas
    §12 Dante organizes 3-sphere slices
    §13 Comments

Chapter 3. Einstein and the 3-Sphere (6 illustrations) 25
    §14 Imagination boggles
    §15 Projecting S^1
    §16 Projecting S^2
    §17 Einstein's view
    §18 Comments

Chapter 4. Einstein's Universe (7 illustrations) 35
    §19 Hollow pipe
    §20 $R \times S^2$ as a solid pipe?
    §21 $R \times S^3$
    §22 Comments

Chapter 5. Images of S^1 and S^2 (24 illustrations).................... 41
    §23 Picturing a one-sphere
    §24 Two-web graph paper
    §25 Picturing a two-sphere
    §26 Three-web graph paper
    §27 Comments

Chapter 6. Four-web Graph Paper (18 illustrations)...................... 53
    §28 What is a hypercube?
    §29 Why is it that we cannot see a hypercube?
    §30 The 4-web and 4-web grids
    §31 Moving the 4-web into 3-space
    §32 Comments

Chapter 7. The Partial Picture (8 illustrations)........................... 65
    §33 Camera position
    §34 First picture
    §35 Walking around our 3-sphere image
    §36 Comments

Chapter 8. Generating the Hyper-Sphere Art (3 illustrations)............. 73
    §37 Overview of chapter
    §38 4-Web grids
    §39 Consistency of spheres
    §40 Points of intersection
    §41 From hyper-space into human view
    §42 Comments

Chapter 9. Prelude to Chapters 10 and 11 (3 illustrations)................ 85
    §43 The fresco, Creation of Adam
    §44 Comments

Chapter 10. Great 2-spheres (13 illustrations)........................... 91
    §45 Three great circles on a 2-sphere
    §46 Four great 2-spheres
    §47 The linear transformation L
    §48 Planar example
    §49 Great 2-sphere images
    §50 The $w = 1/4$ great 2-sphere
    §51 The $z = 1/4$ great 2-sphere

§52 The $y = 1/4$ and $x = 1/4$ great 2-spheres
§53 Comments

Chapter 11. Images of Great 2-spheres (9 illustrations) 109
§54 The $w = 1/4$ case
§55 The $z = 1/4$ case
§56 The $y = 1/4$ case
§57 The $x = 1/4$ case
§58 Topological images
§59 Images of joins
§60 Lower-dimensional example
§61 Joins of great 2-spheres
§62 Comments

Chapter 12. Appendix 1: Supplement for Chapters 1 and 2
    (8 illustrations) 121
§A1 Flexing a disc
§A2 3-disc flex step
§A3 Using a mirror to glue discs
§A4 Patterns and the 3-sphere
§A5 Slices of the 3-sphere
§A6 Hyperplane slices of the 3-sphere
§A7 Dante's 3-sphere construct

Chapter 13. Appendix 2: Supplement for Chapters 3 and 4
    (6 illustrations) 131
§A8 Non-Euclidean geometry
§A9 Context of Einstein's quotes
§A10 Doppler shift
§A11 Red shift
§A12 Big Bang
§A13 Cosmic Background Radiation
§A14 Concerns about Einstein's Universe
§A15 Chronometric Cosmology
§A16 Final paragraph of quote
§A17 Parallel universes and God the observer
§A18 Universal universe

Chapter 14. Appendix 3: Inside S^3 and Questions (4 illustrations) 163
§A19 Inside a 3-sphere
§A20 Physics without measurements?

Chapter 15. Appendix 4: Mathematics and Art (1 illustration)............ 171

Bibliography... 175

Index.. 179

Introduction

Like a coin presents two faces, this book presents two "faces" — a "mathematical face" and an "artistic face". The interplay between math and art is historical (Appendix 4 *Mathematics and Art*); and in this book "mathematics" yields "new art".

The new mathematics relies on the 4-*web*, which may be motivated by thinking of spider webs. Cast in this context, simple pictures allow even non-mathematicians to understand its classical roots (circa 1900) — the *Sierpiński triangle* (2-web) and *Sierpiński cheese* (3-web). The 2-web provides 2-*web art* (page 44), and the 3-web provides 3-*web art* (page 47).

Originally the 4-web lived only in the 4th dimension, and its structure seemed obscure. But by 2003 the 4-web, as well as its global structure, was moved into human view. Within human view the 4-web, like its ancestors the 2-web and the 3-web, yields new 21st-Century 4-*web art* (page 61).

This book tells the rest of the story. Like a spider's web captures insects, the 4-web living in the 4th dimension can capture points of objects, objects that live in the 4th dimension. And when "carried" by the 4-web into human view, the captured points organize into new art. *We can see partial images of objects that live only in the 4th dimension.*

In this book the focus is the new art generated by a *hyper-sphere*.

But what is a hyper-sphere? A hyper-sphere, also known as a 3-*sphere*, is an offspring of an ordinary 2-sphere — we see a 2-sphere each time we look at a full moon or the surface of a basketball. Unlike the surface of a basketball, however, an understanding of the 3-sphere and its *shape* provides a gateway into man's historical thinking about answers to "awe and wonder" questions, questions whose answers lie at the heart of physics and religion:

> What is the shape of our universe?
> How can we picture the Christian afterlife?
> What higher-dimensional shape could produce an image of man?

To become comfortable with the idea of a 3-sphere it is important to know why we cannot see a 3-sphere. Our approach is that of reasoning by analogy: In the graphic below our friend *Freddy the penguin*, who will help us throughout this book, has only 2-*dimensional vision*. It is easy to see why Freddy cannot see a 2-sphere (shape of a basketball).

Freddy cannot see a basketball because he can see neither *above* nor *below* his viewing plane. Freddy needs to upgrade his viewing plane with an *up-down dimension*. Similarly, for humans to see a 3-sphere we would need to upgrade our vision with an *outside-our-physical-space dimension*.

But how can we think about a 3-sphere? We learn to think about a 3-sphere by first understanding that a sphere induces two simple properties — *slicing* and *gluing*. For example, in the illustration above Freddy's viewing plane *slices* the sphere at its equator, thereby dividing the sphere into two hemispheres. And in reverse, we may then recover the entire sphere by gluing its two hemispheres along the equator. Simply put, supported by a sufficient number of pictures, the reader will realize that it is the slice and glue properties that underlie our understanding of a 3-sphere.

With the idea of slicing and gluing in hand, we illustrate how Dante, in his world-famous *Divine Comedy*, used a mirror to describe a gluing that yielded a 3-sphere, i.e., Chapter 2 contains numerous pictures that guide the reader to an understanding of how a 3-sphere is used to allegorically picture the Christian afterlife.

Moving to answer the first "awe and wonder" question, we have Chapter 3 *Einstein and the 3-sphere* where we quote Einstein's description of a 3-sphere. But where Einstein used only one picture, we provide numerous pictures with elucidating comments. Each graphic adds clarification. Likewise, Chapter 4 *Einstein's Universe* (EU) also contains many illustrations that demonstrate how EU tells us that at any instant in *time* our universe is a 3-sphere.

And so after having shown how the 3-sphere has served as a foundation on which to build answers to the first two "awe and wonder" questions, the second half of the book is devoted to using the 3-sphere to build a foundation for answering the last "awe and wonder" question, namely,

What higher-dimensional shape could produce an image of man?

In particular, the *self-similar 4-web fractal* inside the 4th dimension induces *4-web grids* that contain millions of increasingly-smaller triangles. These triangles are used to capture literally tens-of-thousands of points on a 3-sphere, which are then moved into human view. The result is the new art pictured on the cover. But perhaps more importantly, because the new *capture and move into human view* method is general, it may be applied to other structures that live only in the 4th dimension.

This *new art* technique is illustrated and explained in detail. For example, Freddy the penguin shows us, by way of analogy, how the captured points may be moved into human view (§31). In addition, the first movie on the supplemental Blu-ray Disc shows a *level-1 4-web grid moving into human view*.

After the hyper-sphere graphic is within human view, the image is discussed from various perspectives. The upshot is that a 3-sphere produces an image that many view as the *face of man* — a 3-sphere provides an answer to the third "awe and wonder" question. To see the depth of the hyper-sphere graphic, five videos are provided on a supplemental Blu-ray Disc.

This book Incipit (here begins) is the first-ever presentation of the new art technique for picturing objects that live in the 4th dimension. In technical terminology, the artist's brush is the computer, and the brush strokes that produce the new art are guided by an algorithm based on the 4-web *self-similar fractal*. For those who desire to apply this new method to create their own "new art", Chapter 8 provides not only the line-by-line computer code that was used to generate the hyper-sphere graphic, but also many comments that add insight into the design of the code. Chapters 9, 10, and 11 serve to motivate, develop, and graphically specify certain *great* 2-*spheres* within a hyper-sphere.

The 4-web may well be the simplest self-similar fractal to create such art because its birth was conceived by selecting *a single point in the 4th dimension*. To illustrate, suppose a spider from the 4th dimension decides to weave a web (see graphic below). He begins by weaving a triangle, i.e., a *level*-0 2-*web*. Then moving above the plane of the 2-web, he selects a single point "*p*" and weaves segments from *p* to the corners of his 2-web. He now has a *level*-0 3-*web* (right-side graphic). He then *judicially selects* a single point *q in the 4th dimension* and weaves segments from *q* to the corners of his 3-web, thereby creating a *level*-0 4-*web*. *Why is the point q not pictured?*

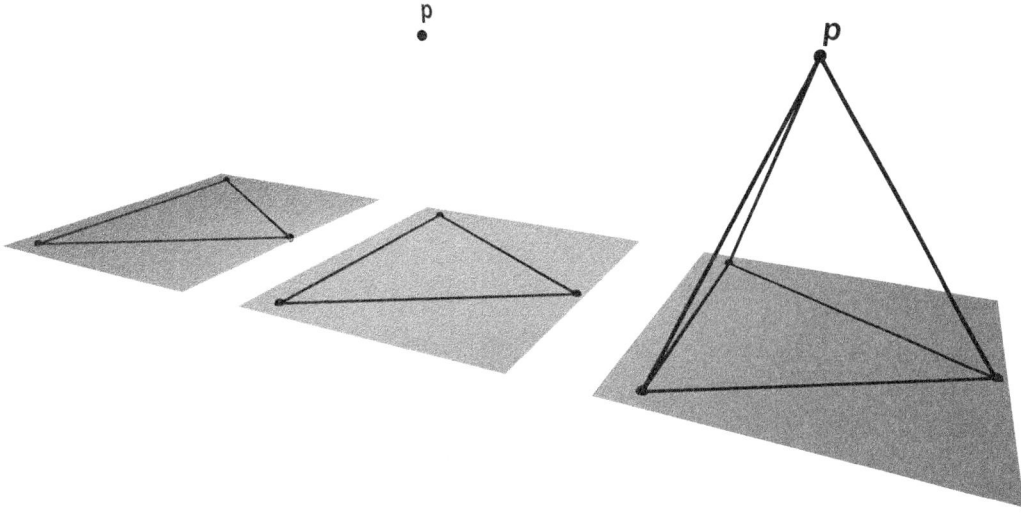

CHAPTER 1

3-Sphere

Since the beginning of human civilization, man has struggled to understand his universe. Ironically, the Aristotle (384-322 BC) spherical model, with earth at the center, was fundamental to Dante's (1265-1321 AD) model which, in turn, underlies the so-called Einstein (1879-1955 AD) Universe. In common language, the Aristotle model involves a solid ball with a spherical surface, and Dante's model adds a mirror-image solid ball to Aristotle's and then joins them by "gluing their surfaces." In 20th Century language, Dante constructed a three-sphere S^3; and the three sphere S^3 turned out to be fundamental to what is commonly called "Einstein's Universe". All of these models involve *spheres* constructed inside universes of dimensions three and four.

But what does it mean to say *spheres constructed inside universes of dimensions three and four*? And how can we think about and understand such spheres without getting bogged down in the mathematical jargon? In this chapter we answer these questions by using simple examples and pictures.

§1 SPHERES

We focus on four types of spheres. Three of the four types are pictured in Figure 1.1 below — from left-to-right we see models of a 0-sphere (two points), a 1-sphere (circle), and a 2-sphere (surface of a ball which may be viewed as a seamless inflated balloon):

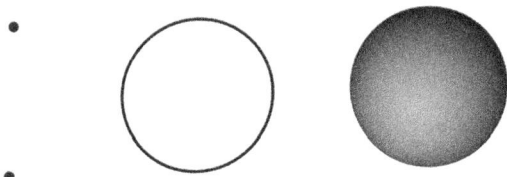

FIG. 1.1 A zero-sphere S^0, a one-sphere S^1, and a two-sphere S^2.

The fourth kind of sphere is the 3-sphere S^3. Unlike S^0, S^1, and S^2, we cannot picture S^3 because human vision is 3-dimensional — it is impossible to picture S^3 within such a "dimension-restricted" visual universe.

But what does it mean to say 3-*dimensional*? Historically, a precise definition of dimension did not appear until the early 1900s, and the definition involves some subtle mathematics. So we shall take another approach, namely, in the following few pages we shall use examples and pictures to convey the essence of "3-dimensional."

§2 WHY IS IT THAT WE CANNOT SEE A 3-SPHERE?

To roughly appreciate why human vision does not allow us to picture S^3, consider the following S^2 analogy: On the left side of Figure 1.2 we see Freddy the penguin with "2-dimensional vision" looking at S^2 (say the surface of a basketball). But because Freddy only has 2-dimensional vision — he only sees what lies within the "triangular plane issuing from his eyes." He cannot "see" the "roundness" of the basketball, but he does see the part of the basketball that intersects his "triangular plane of vision." On the right side of Figure 1.2 we see the circle that Freddy sees, but Freddy sees the circle only from its edge.

FIG. 1.2 Freddy, with 2-dimensional vision, looks at a 2-sphere.

Freddy cannot see S^2 because S^2 contains points above and below his viewing plane. And to move off of the viewing plane, say in an "up-down direction" requires an extra dimension. Freddy's vision, unlike human vision, does not include the extra "up-down direction." So Freddy cannot see the 2-sphere S^2, while humans easily see S^2.

Intuitively at least, this example with Freddy motivates the belief that humans (with 3-dimensional vision) cannot see the 3-sphere S^3. Just as S^2 "curves" outside of any 2-dimensional plane, S^3 "curves" outside any 3-dimensional *hyper*plane.

Other facts illustrated in Figure 1.2 extend to the 3-sphere S^3. Most notably, when Freddy's visual plane intersects S^2 as indicated, he sees S^1. Step this experiment up one dimension, and we have the fact that *when humans similarly look at S^3, they see S^2*.

In particular, when we look into the night sky and see a full moon, we may be seeing that part of a 3-sphere where it intersects our visual universe. (See Figure 1.3 where Freddy has recovered his 3-dimensional vision.)

FIG. 1.3 "Is it an S^3 intersecting my 3-dim visual plane?"

§3 SPHERES AND DISCS

Potatoes have skins, boiled eggs have shells, and planets have surfaces. *But what does a potato, a boiled egg, and a planet illustrate?* Each provides an intuitive example of a "solid ball" bounded by a "2-sphere." To be sure, the bounding 2-spheres are represented by the skin, the shell, and the surface. We shall call a solid ball that contains its 2-sphere boundary a 3-*disc*.[1]

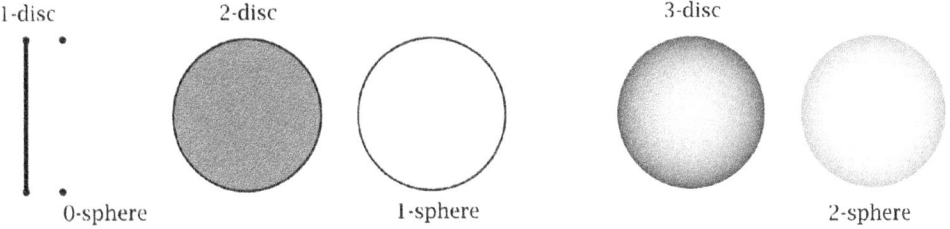

FIG. 1.4 Discs with their spherical boundaries separately illustrated.

Let us step down to dimension two: Each flat side of a coin has a unique edge, an archery target has an area inside of a circle, and within a picture of a wheel we see an area bounded by a rim. At dimension two these are rough examples of a "circular area" bounded by a "circle" (one-sphere S^1). We shall refer to an area that contains its bounding circle (S^1) as a 2-*disc*.

---

[1]The "3" prefix in "3-disc" tells us that a 3-disc has three dimensions (the number of dimensions required for volume), just as the "2" prefix in "2-sphere" tells us that a 2-sphere has two dimensions (the number of dimensions required for area). Similar interpretations may be applied to the prefixes in 2-disc, 1-disc, 1-sphere, and 0-sphere.

Now let us step down to dimension one: A needle has a sharp endpoint and a dull endpoint, a carpenter marks a line between two points, and a single human hair has a root and a tip. At dimension one these statements provide rough examples of a "line segment" bounded by "two endpoints" (zero-sphere S^0). We shall call a line segment that contains its bounding zero sphere S^0 a 1-*disc*.

§4 Gluing discs

Lines and line segments are the simplest examples of one-dimensional objects — a *line* ℓ and a 1-disc (line segment) $[p,q]$ with *endpoints* p and q are pictured below where the dots "$\cdots$" indicate that the line extends indefinitely. A line "does not stop," it has infinite length. In contrast, a line segment $[p,q]$ is the "connected part" of a line that lies between, and includes, its two distinct endpoints p and q. A line segment has finite length.

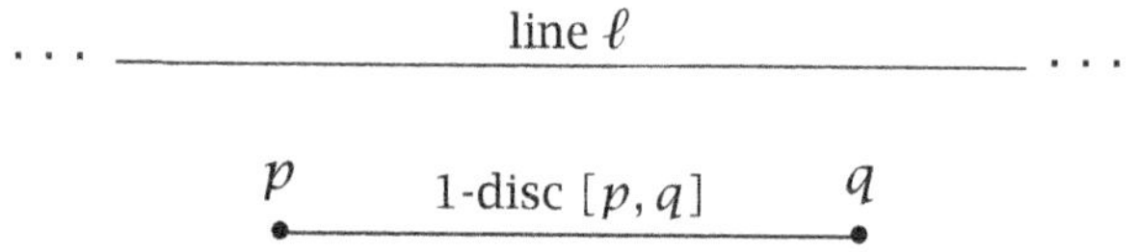

$\cdots$ ——————————— line ℓ ——————————— $\cdots$

p ●————— 1-disc $[p,q]$ —————● q

Since the end points of a 1-disc are distinct, they are points of a 0-sphere — *the boundary of a 1-disc is a 0-sphere S^0*.

We can "glue" two 1-discs along their S^0 boundaries to obtain a one-sphere S^1 (Figure 1.5): Starting with two 1-discs, say $[p,q]$ and $[p',q']$, we first "flex" each of these discs into a semicircle, then we join the two semicircles by "gluing" p to p', and, q to q'. After the gluing we have $p = p'$ and $q = q'$ as illustrated, and the result is a one-sphere S^1.

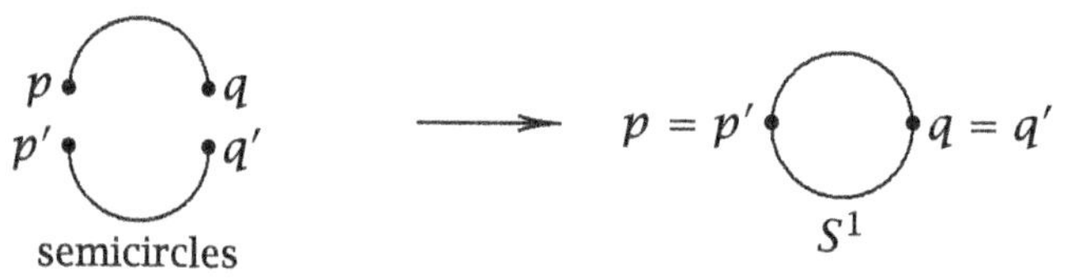

Fig. 1.5 Two 1-discs glued at their S^0 boundaries yield S^1.

Next, we step up to the two-dimensional case. The simplest examples of two-dimensional objects are planes and either "curved" or "flat" surfaces:

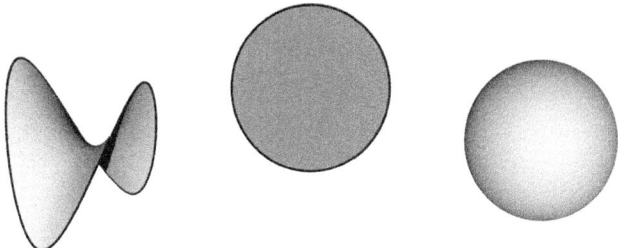

From left-to-right we see a *saddle surface* (horse saddle), a 2-disc (circular flat area bounded by a circle S^1), and a 2-sphere (seamless inflated balloon).

We can also "glue" two 2-discs along their S^1 boundaries to obtain a 2-sphere S^2 (Figure 1.6): Starting with two 2-discs, we first "flex" each of these discs into a hemisphere, then we join the two hemispheres by "gluing their S^1 boundaries." After the gluing, the two S^1 boundaries are equal as illustrated, and the result is a two-sphere S^2.

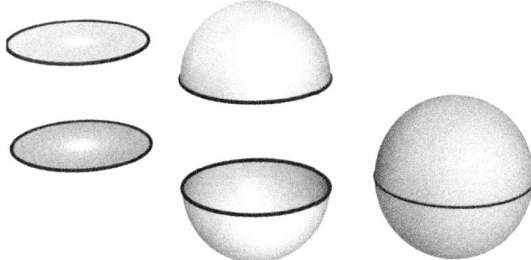

FIG. 1.6 Two 2-discs glued at their S^1 boundaries yield S^2.

Gluing boundaries of discs, as detailed above for the 1-disc and 2-disc cases, extends to the 3-disc. The 3-disc case yields a 3-sphere S^3:

S^3 *is the object obtained by gluing two 3-discs along their S^2 boundaries.*

Unfortunately, however, as discussed in §2 above, S^3 is not visible. But let us return to the question, *Why is it that we cannot see a 3-sphere?* We begin with Freddy and a few scientists conducting experiments (Fig. 1.7).

FIG. 1.7 Freddy is standing still and concentrating.

With two-dimensional vision on the left, and one-dimensional vision on the right, Freddy is thinking, *I see nothing moving. Why did these scientists tell me to stand still and concentrate?* Like magic — while Freddy sees nothing change — the insides of both discs *flex* into an extra dimension (Figure 1.8).

FIG. 1.8 Freddy cannot see the "insides" flex into another dimension.

And after the experiment is over, the scientists ask Freddy, *Did you see anything change?* Freddy replies, *Nothing changed!* Freddy then asks, *What is the purpose behind asking me to look at the discs?* They responded with the following:

> We wanted you to experience a pattern: With 1-dimensional vision you could not see a 1-disc flex into an extra dimension. With 2-dimensional vision you could not see a 2-disc flex into an extra dimension. So based on these experiments, the general pattern is the following: *With __-dimensional vision you cannot see a __-disc flex into an extra dimension.* The idea is that you can fill in both blanks with the number "3".

Freddy responds, *Oh I see, "3" placed in the blanks implies humans with their 3-dimensional vision cannot see a 3-disc flex into an extra dimension.*

You got it. And if humans cannot see a 3-disc flex into 4-dimensional space, you might ask yourself, *Can humans see a 3-sphere?* Freddy pauses, and then says, *I would guess not, because how would they see the gluing of two flexed 3-discs if they could not see one flexed 3-disc?*.

You got it! Human visual sensors (their eyes) are not designed to "see" four dimensions. But humans do use their mind's eye to see that the mathematics of "flexing a 3-disc" is essentially the same as the 2-disc and 1-disc cases. In reality, we need "God like" vision — at least 4-dimensional vision — to "see" a 3-sphere.

§5 Slicing spheres

We can think of a sphere like we think of a loaf of bread — we can slice the loaf to obtain slices, mix up the slices, then solve the puzzle of reorganizing the slices so that we can see the original loaf.

And just as a loaf of bread contains many slices, a one-sphere S^1 contains many zero-spheres (Figure 1.9): Given a one-sphere S^1 (a circle), we may "slice" S^1 with a vertical line — similar to slicing a loaf of bread with a knife. Each "slice" yields either two points p and q (a 0-sphere) or a single point r.

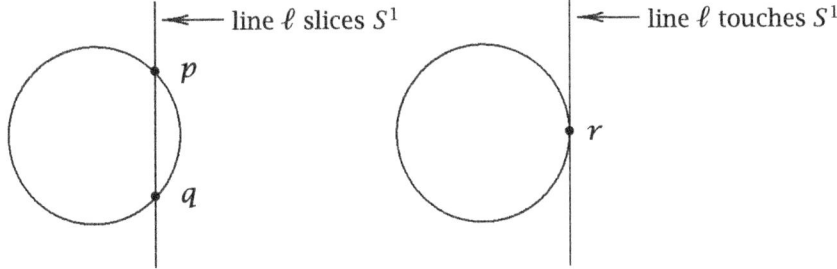

FIG. 1.9 The 1-sphere consists of 0-spheres and two points.

Focus on the left-side circle (1-sphere). If we move a vertical line from left-to-right across the 1-sphere, then, except for the initial and terminal vertical lines, each slice yields a 0-sphere (two points). The left-side graphic shows the moving line at an instant after it has moved past about 75% of S^1, while the right-side graphic illustrates the "right-side" terminal line.

We may also ask Freddy, with his one-dimensional vision, to illustrate slicing of a one-sphere (Figure 1.10).

Freddy sees a "slice" of a one-sphere as a *single* point. Note that as Freddy looks at the slice in the right-side illustration, two points are within his vision line. But the point closest to Freddy blocks his view of the other point.

FIG. 1.10 Freddy's 1-dimensional vision slices the circle S^1.

Stepping up one dimension, we now consider the 2-sphere. Our slicing knife must change — we move from a one-dimensional line to a two-dimensional plane. The idea is pictured in Figure 1.11.

FIG. 1.11 The 2-sphere consists of 1-spheres and two points.

Focus on the 2-sphere on the left side. If we move a vertical plane from left-to-right across the 2-sphere, then, except for the initial and terminal vertical planes, each slice yields a circle (1-sphere). The left-side graphic shows the moving plane at an instant after it has moved past about 75% of S^2, while the right-side graphic illustrates the terminal plane.

As for slicing a 3-sphere, we again change the knife. This time from a 2-dimensional plane to a 3-dimensional *hyper*plane. In this case the result is analogous to those for the one-sphere and two-sphere.

§6 COMMENTS

Within this chapter we developed an intuitive understanding of the zero-sphere S^0, one-sphere S^1, two-sphere S^2, and three-sphere S^3. Since it is impossible for us to literally see a three-sphere S^3, we studied properties of the spheres of dimension less than three that generalize to S^3.

For example, we constructed a one-sphere from two 1-discs by gluing their zero-sphere boundaries. And we constructed a two-sphere from two 2-discs by gluing their one-sphere boundaries. This pattern continues and allows us to construct a three-sphere S^3 from two 3-discs by gluing their two-sphere boundaries. From the *dimension perspective* the method of positioning and gluing discs to create spheres appears in Appendix 1 *Supplement for Chapters* 1 *and* 2.

For another example, we saw how a line may "slice" S^1, and then how a plane may "slice" S^2. And while we cannot "see" the 4th-dimension, our visual space is a *hyperplane*, i.e., an extension of an ordinary plane, and a hyperplane similarly slices S^3 into slices, each slice being either a 2-sphere or a single point. A proof is provided in Appendix 1 §A6 *Hyperplane slices of a* 3-*sphere*.

To complete this chapter we would be remiss if we did not recall *the standard approach to defining spheres*: Select a *space* — 1-*space* (a line), 2-*space* (a plane), 3-*space* (our 3-dimensional human visional *hyper*plane), or 4-*space* (3-space plus one extra dimension). The extra dimension allows us to move outside of our 3-space. Next, select a *center-point* "c" in the "selected space" and a positive *radius-number* "r". Then within the "selected space," *a sphere with center c of radius r* consists of all points whose distance from the center c is r. For example, if we select 4-space, then the standard approach describes a 3-sphere S^3 centered at c of radius r.

For those who may recall their high-school mathematics of the x, (x, y), (x, y, z), and (x, y, z, t) coordinate systems for a line, a plane, 3-space, and 4-space, respectively, also recall the following equations where the center c is the "origin" of the "selected" coordinate system:

$S^0 : x^2 = r^2$ (0-sphere: two points $x = r$ or $x = -r$ on a line)
$S^1 : x^2 + y^2 = r^2$ (circle in an x-y plane)
$S^2 : x^2 + y^2 + z^2 = r^2$ (surface of seamless inflated balloon in 3-space)
$S^3 : x^2 + y^2 + z^2 + t^2 = r^2$ (3-sphere in 4-space)

Finally, the author enjoyed being one of the parents of Freddy the Penguin. The other parent is Robert McNeel & Associates who created their *Model of a Toy Penguin* using their *Rhinoceros* 3-D modeling software, which is detailed in their *Users Guide* Version 3.0.

Freddy inherited his eyes and gray-scale shading from my side of the family, but his good looks and excellent body structure came straight from the McNeel side.

CHAPTER 2

Dante's 3-Sphere Universe

Dante Alighieri (1265–1321) AD wrote what is now considered one of the greatest works of world literature, the *Divine Comedy* — an allegorical vision of Christian afterlife.[1] The focus here is a description of a universe that includes the *Empyrean*, which, among Christian poets, is the abode of God or the firmament. Dante constructs the Empyrean as a mirror image of the classical Aristotle universe, and then "glues their 2-sphere boundaries" to form a 3-sphere. We essentially follow the article *Dante and the 3-sphere* by Mark Peterson, American Journal of Physics 47 (1979).

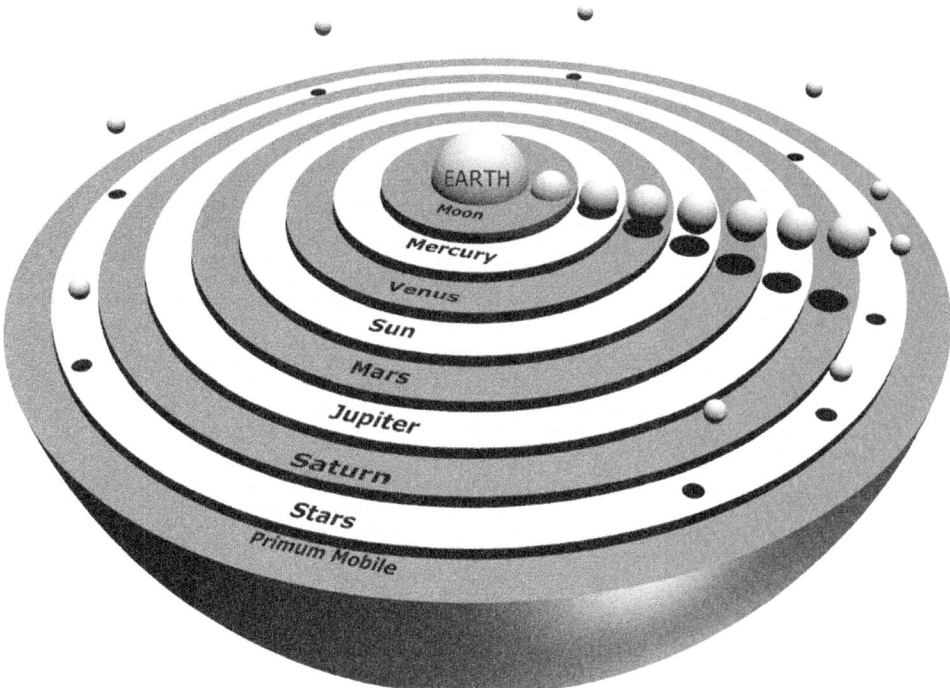

FIG. 2.1 *Heavenly sphere* — classical Aristotle/Greek model of the universe.

---

[1] An *allegory* is a description of one thing under the image of another.

§7 ARISTOTLE UNIVERSE

Underlying Dante's Divine Comedy is the Aristotle (Greek) model of our universe. The illustration above enhances the following quote from John Stillwell's *Yearning for the Impossible* published by A. K. Peters, Ltd., 2006 (Figure 2.1).

> [*Page 102 of Stillwell's book; square brackets "[·]" indicate my comments.*]
> The Greeks believed the universe should reflect the geometric perfection of circles and spheres, and they imagined space structured by a system of spheres. ... The earth is the innermost sphere, surrounded by eight concentric "heavenly" spheres carrying the known celestial bodies, and an outermost sphere called the Primum Mobile [in the Ptolemaic system, the tenth and outermost concentric sphere, revolving from east to west around the earth and causing all celestial bodies to revolve with it]. (For example, "seventh heaven" is the sphere of Saturn.) The motion of the sun, moon, planets, and stars was attributed to the rotation of the spheres carrying them, with the Primum Mobile ("first mover") controlling them all. Somehow, the ancient universe stopped at the Primum Mobile, ...

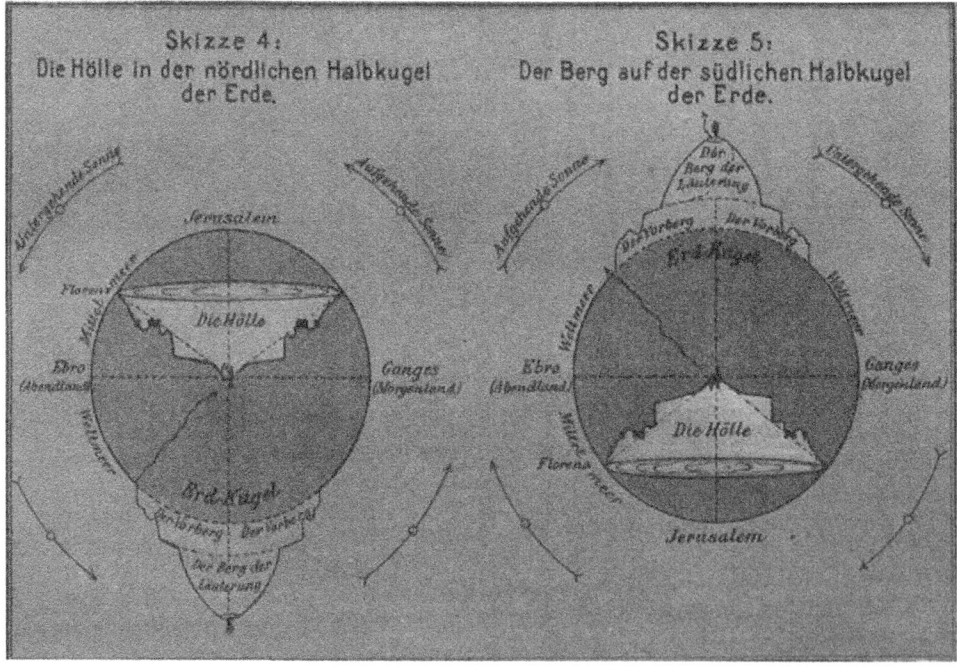

FIG. 2.2 Albert Ritter's sketches of Comedy's geography.

§8 Dante's journey

Centered around Dante's afterlife journey (the journey of his Christian soul) into Heaven, the poem is divided into three major parts, the *Inferno*, *Purgatorio*, and *Paradiso*. In *Inferno* we have the beginning of his journey where Dante descends into Hell. Satan's fall created Hell as a funnel-shaped gateway (rock displacement) under Jerusalem that leads to the center of the earth. The rock displacement in turn created *Mount Purgatory* on the surface of the earth diametrically opposite to Jerusalem (Figure 2.2).

The sphere on the right in Figure 2.2 is an inversion of the one on the left. Starting at Florence, located on the left side of the "funnel to Hell" in the left-side sphere in Figure 2.2, Dante, moving toward the center of the earth, descends into Hell. He then continues down, and thereby up in the right-side sphere, to Mount Purgatory's shores in the southern hemisphere. Then he ascends to the top of Mount Purgatory, the location of the *first sphere of Heaven* — the "ANGELS" part of the *Angelic sphere* (Figure 2.3).

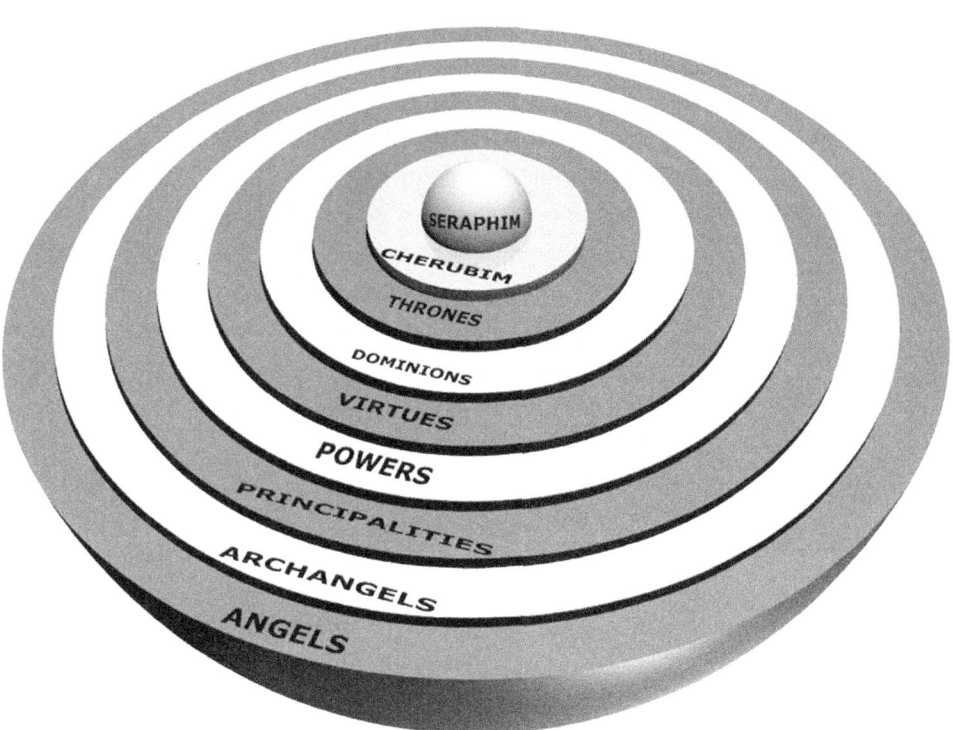

FIG. 2.3 *Angelic sphere* (Empyrean: home of God and Angels).

As Dante descends into Hell, his soul sees sin for what it really is — self-indulgent sins, violent sins, and malicious sins. But escorted by an Angel, Dante's soul survives Hell, and then Dante and the Angel move from the center of the Earth to Mount Purgatory. *Purgatorio* allegorically represents the Christian life — angels bring Christian souls for conversion from sin to the state of grace.

In Paradiso, we find Dante after he leaves the Aristotle universe and enters "Heaven" which is the *Empyrean* (among Christian poets, the abode of God, the firmament).

§9 Angelic sphere

The *Angelic Sphere* is a mirror image of the Aristotle Universe. The "spheres of angels" or *Angelic Choirs* illustrate an ordering relative to "closeness to God," who resides at the center of the center sphere.

> *[Page 104 of Stillwell's book; square brackets "[·]" indicate my comments.]*
> In Canto [main division of poem, analogous to "chapter" division of a book of prose] XXVIII Dante views the Empyrean as not only the *complement* but also the *reflection* of the heavens visible from earth. He makes a smooth transition from the heavens to the Empyrean by using the Primum Mobile as a half-way stage between two worlds, the "model" and the "copy." From this vantage point, he sees the heavenly spheres on one side as an image of the angelic spheres on the other.
>
> > *as one who in a mirror catches sight*
> > *of candlelight aglow behind his back*
> > *before he sees it or expects it,*
> >
> > *and, turning from the looking-glass to test*
> > *the truth of it, he sees that glass and flame*
> > *are in accord as notes to music's beat*[2]

FIG. 2.4 Dante and Beatrice gaze upon the Empyrean.[2]

We continue with our quote from Stillwell's book:

> With this sophisticated model of a finite universe, the Church was able to hold out against infinite space for a few centuries. But eventually infinite flat space came to be generally accepted for its greater simplicity, despite some uneasiness about infinity ...
>
> In the twentieth century, cosmology returned to the idea of a finite universe, and physicists now look back in admiration to Dante's *Paradiso*, seeing in it a good description of the simplest finite universe, which we now call the 3-*sphere*

---

[2]Translation by Mark Musa of line 4-9, Canto XXVIII, of Dante's Paradiso. Art by Gustave Doré (1832-1883).

§10 Dante's 3-sphere

To reinforce Dante's construction of his universe let us consider the Aristotle Universe, abstractly a 3-*disc* (solid ball whose boundary is a 2-sphere), and its mirror image the Empyrean, another 3-disc (Figure 2.5).

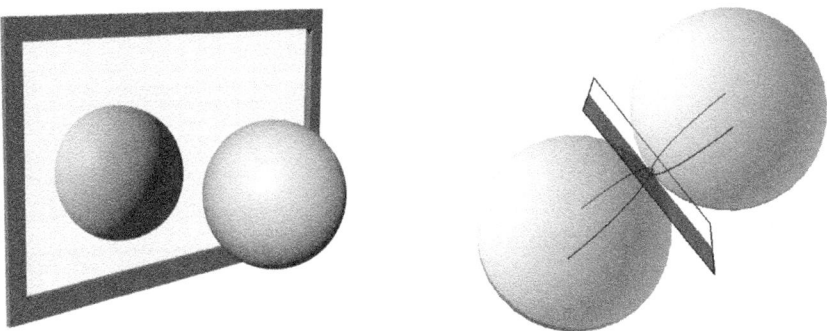

FIG. 2.5 Aristotle Universe just touching its mirror image the Empyrean.[3]

On the right side of Figure 2.5 we see how Dante views the gluing of *a single point* on the boundary of the Aristotle Universe with *a single point* on the boundary of the Empyrean. The single point (where the two spheres meet) may also be viewed — within Gustave Doré's art in Figure 2.4 — as the location of Dante and (the angel) Beatrice with the Empyrean in the background.

Dante's construct implies that each point on the boundary of the Aristotle Universe corresponds to one and only one point on the boundary of the Empyrean, and conversely, each point on the boundary of the Empyrean corresponds to one and only one point on the boundary of the Aristotle Universe. That is, using "Dante's mirror," we see that as the sphere in the left-side graphic in Figure 2.5 rotates, its mirror image would likewise rotate, yielding a *faithful matching* between the boundaries of the Aristotle Universe and the Empyrean.

It follows that Dante *faithfully glues the S^2-boundary of the Aristotle Universe to the S^2-boundary of the Empyrean*. Dante is constructing a 3-sphere.

One benefit of Dante's construction is that of removing the "edge" of the Aristotle Universe.

> [*Mark Peterson - Dante and the 3-sphere*] The belief that the earth must be round goes back at least to Aristotle, whose doctrine of "natural place" required a round earth at the center of the universe. This same model became central to Christian theology with the work of Thomas Aquinas, and it forms the cosmological framework for Dante's *Divine Comedy*.

---

[3]The "mirror graphic" on the left was provided by Mary Fugier, who used Rhinoceros 3-D Software (www.rhino3d.com).

> The belief that the universe as a whole might be round (or more generally, curved) is a much more recent one. It seems to require mathematics of the 19th century (non-Euclidean geometry) even to formulate the notion.
>
> It is therefore a considerable surprise to find, on closer reading, that Dante's cosmology is not as simple geometrically as it at first appears, but actually seems to be a so-called "closed" universe, the 3-sphere, a universe which also emerges as a cosmological solution of Einstein's equations in general relativity theory.
>
> I came upon this suggestion about Dante and the 3-sphere in wondering how Dante would treat an evidently unsatisfactory feature of the Aristotelian cosmology when he, as narrator in the *Paradiso*, got to the "edge" or "top" of the universe. How would he describe the edge? It is the same problem every child has wondered about: unless the universe is infinite, it must (the argument goes) have an edge — but then what is beyond? Dante faces this very problem at the end of the *Divine Comedy* where he must describe the Empyrean not in terms of principles or abstractions, as the standard cosmology did, but as someone actually there.

Within the last four words of Peterson's quote, namely, *as someone actually there*, the last word *there* means *at the boundary of the Aristotle Universe* where the gluing takes place. The "there" may be graphically viewed as the point in the right-side illustration of Figure 2.5 where the two spheres are just touching.

But how can we picture the "locally glued area" near the "there" point?

§11 LOCALLY GLUED AREAS

Referring again to Gustave Doré's art in Figure 2.4, we see Dante and Beatrice standing at the *there* point where we also see "land" or a "local area" on Mount Purgatory that surrounds Dante and Beatrice.

Expressed differently, if as illustrated on the right side of Figure 2.5 we select a point p on the 2-sphere boundary of the Aristotle Universe that we shall glue to a point p' on the 2-sphere boundary of the Empyrean, then can we view an *area of glued points* that are "close" to a single glued point $p = p'$?

The gluing process is that of gluing *points near p* to *points near p'*. But as illustrated on the right side of Figure 2.5, we see how the 2-sphere containing p curves away from p and how the 2-sphere that contains p' curves away from p'. So it is not immediately obvious how we could picture such a "local gluing."

With Figures 2.6 and 2.7 we nevertheless provide a way to view a *local gluing*. Beginning with the left-side of Figure 2.6, we locally-flatten two

2-spheres. Then we *faithfully match* the points within the small white area with their counterpart points on the adjacent flattened sphere — we glue (right-side illustration) the locally-matched areas.

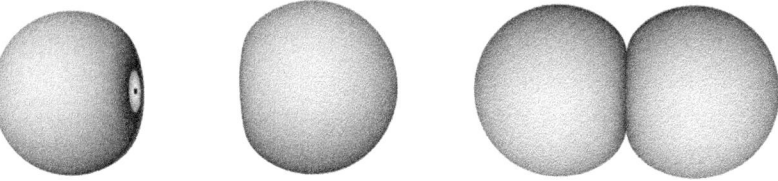

FIG. 2.6 We locally flatten two 2-spheres, and then glue locally-matched areas.

By making one of the flattened spheres transparent, we see the *locally glued area* illustrated as the relatively-white circular area in Figure 2.7.

FIG. 2.7 The white area near the black dot is a "locally glued area."

Using Figure 2.7 we may make another observation. If the black dot is viewed as the location of Dante and Beatrice in Figure 2.4, then in reverse, the area of land surrounding Dante and Beatrice corresponds to the white circular area surrounding the black dot.

Because every point on the 2-sphere boundary of the Aristotle Universe is surrounded by a "locally glued area," we may reason that one may *cross over* from the Aristotle Universe to the Empyrean at every "glued point."

This observation shows that Dante's Universe has *no edge*, as Peterson implies in the last paragraph of his quote (page 17).

Keep in mind that the 3-sphere structure is not unlike its lower-dimensional analogues — the 1-sphere has two *semicircles* that meet in a 0-sphere (Figure 1.5); the 2-sphere has two *hemispheres* that meet in a 1-sphere (Figure 1.6); and, likewise, the 3-sphere has two *hyperhemispheres* — the Aristotle universe and the Empyrean — that meet in a 2-sphere. The *equator* of the 3-sphere is a 2-sphere, and the 3-sphere has no edge.

§12 DANTE ORGANIZES 3-SPHERE SLICES

In Figure 2.8 we slice a 1-sphere S^1 (circle) with vertical lines to obtain 0-spheres, and in Figure 2.9 we slice a 2-sphere S^2 with vertical planes to obtain 1-spheres:

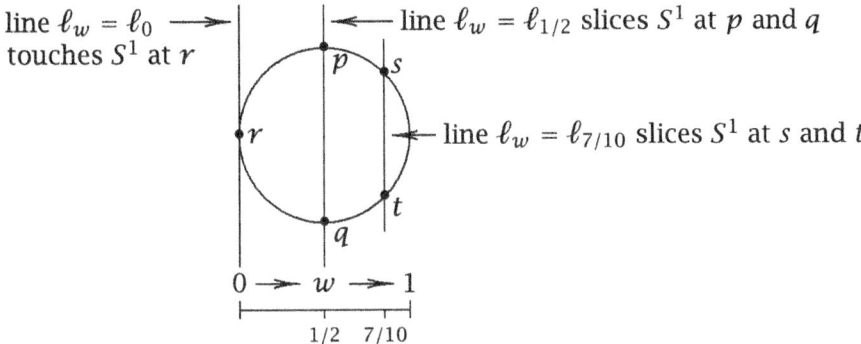

FIG. 2.8 As the values of w increase, the line ℓ_w moves across S^1.

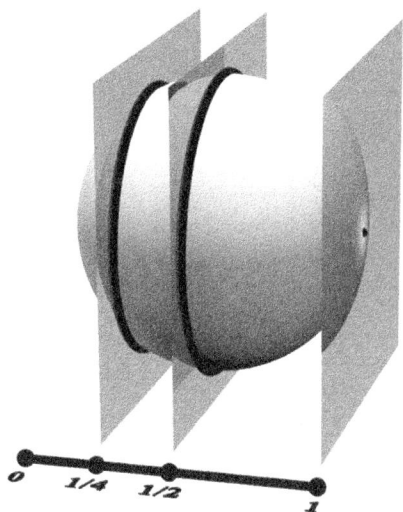

FIG. 2.9 As values of w increase, the plane π_w moves across S^2.

We begin with Figure 2.8: For each number "w" between zero and one, there is a vertical line ℓ_w; and as w moves left to right from zero to one, ℓ_w moves from left to right slicing the circle.

Letting the *size of a 0-sphere be the distance between its two points*, we start on the left with the ℓ_0-slice as a point, then the sizes of the ℓ_w-slices increase to a maximum at $\ell_{1/2}$, and then decrease as ℓ_w moves further to the right until the final ℓ_1-slice is simply a point.

Turning to Figure 2.9, we see the slicing of a 2-sphere with vertical planes that yield 1-spheres slices. Again, as the values of w increase from zero to one, the corresponding planes π_w move across a 2-sphere S^2, yielding slices that are one-spheres — the first π_0-slice is a point, then the sizes (diameters) of the π_w-slices (one-spheres) increase to a maximum at $\pi_{1/2}$, and then the π_w-slices decrease in size as w moves from 1/2 to 1, where we find that the π_1-slice is a point.

Slicing a 3-sphere is similar — using 3-dimensional *hyper*planes (human vision spaces) we obtain slices that are 2-spheres, except for the $w = 0$ slice and the $w = 1$ slice which are points. And even though humans cannot picture a 3-sphere, there are mathematical proofs (see Appendix 1) that show the behavior is exactly as described.[4]

Against this backdrop there emerges another way to view these spheres. Consider the one-sphere S^1 (circle). Rather than starting with the S^1, the *whole loaf of bread*, and then creating the *individual slices*, which are 0-spheres, suppose we reverse the process. That is, suppose we start with a *bunch of slices*, say a bunch of 0-spheres, and then consider *how to put these slices together to obtain the whole loaf S^1*. The key is using the values of the *extra dimension* encoded as "w" in Figure 2.8. To start we could consider the largest $\ell_w = \ell_{1/2}$ slice, and go from there.

How can we make it clearer? Perhaps by saying that the slices must simply be lined up according to the values of w. The first slice is a point — the ℓ_0-slice. The unique largest slice would be the $\ell_{1/2}$ slice, and so on. In summary, we could place the slices according to size until they were lined up as in Figure 2.8. And once together, we would see the entire loaf, i.e., we would see the 1-sphere, not the individual slices.

The key idea is the observation that the 0-spheres expand/contract in a *vertical direction*, while it is the *horizontal direction* induced by w that adds the *extra dimension* that is required if we desire to generate a 1-sphere from 0-spheres.

The idea of *building the whole loaf from its slices using an extra dimension induced by w* lies at the heart of Dante's argument. Dante's approach is discussed within the Peterson quote below, which is consistent with a careful reading and consideration of the examples above. In fact, Dante's introduction of the *extra dimension w* allows him to construct his 3-sphere universe inside 4-dimensional space.

---

[4]The proof presented in Appendix 1 requires some background in either vector analysis or linear algebra. It is included for those readers who desire easy access to such a proof.

§12 DANTE ORGANIZES 3-SPHERE SLICES

[*Mark Peterson - Dante and the 3-sphere*] Dante himself believed he was expressing something entirely new at this juncture. He asserts this by describing the difficulty of the notion as being like a knot that has grown tight, because no one has ever before tried to untie it.[5] There can be little doubt, however, that his new idea had no effect on cosmological thinking whatever — the 3-sphere in the *Paradiso* went unnoticed, or ununderstood. In recent times it has probably been dismissed by readers with less geometrical aptitude than Dante as mysticism.

To make the case, then, I first point out that Dante assumes from the outset that the nine angelic spheres and the nine heavenly spheres are analogous. In the notes to his translation, Ciardi makes this point by describing the angelic spheres as "a sort of counter-universe."[6] In fact, what interests Dante as narrator is a seeming *breakdown* in the analogy, about which Beatrice quickly reassures him.[7] The problem is that the various heavenly spheres revolve faster in proportion as they are bigger, while just the reverse is true of the angelic spheres: the innermost and smallest of these are revolving the fastest, and the outer ones are slower. Beatrice replies that if he will shift his attention away from the spheres' sizes to an intrinsic ranking they possess, he will see a marvelous consistency in the whole. The innermost angelic sphere turns faster than the other angelic spheres because it ranks higher, just as the Primum Mobile turns faster than the other heavenly spheres because it ranks higher. In other words, the spheres have a ranking, a "greatness," which does not necessarily correspond to their size (although for the first nine it does), but is rather indicated to the eye by their speed. This explanation strongly suggests our construction of the 3-sphere as sliced up into 2-spheres which at first grow and then diminish in size, labeled by a fourth coordinate w, which simply increases. Indeed, Dante has actually introduced such a fourth coordinate to label the spheres as they grow and diminish, namely their speed. In all our visualizations of the 3-sphere it was the second hemisphere, composed of the diminishing sequence of 2-spheres, which was hardest to fit into the model — Dante embeds the model in four dimensions, which does, as we know, solve the problem. His fourth dimension is speed of revolution. Of course he would never have said it that way, but it amounts to the same thing. The overall organization of the 2-spheres is that of a 3-sphere.

---

[5] Dante, *Paradiso* (Harvard Univ., Cambridge, 1972), Canto 28, lines 58-60.
[6] John Ciardi in Dante, *The Paradiso* (Mentor, New York, 1970), notes to Canto 28, lines 21-36, p. 313.
[7] Dante, *Paradiso* (Harvard Univ., Cambridge, 1972), Cantor 28, lines 46-78.

Dante's elation with this idea — a feeling we may readily share — has traditionally left readers somewhat puzzled. That is just another way of saying that if this passage is not taken as a description of the organization of 2-spheres into a 3-sphere, then it is hard to see what the point of it is.

§13 COMMENTS

Students of 20th Century mathematics know of the construction of the 3-sphere as the object obtained by gluing the 2-sphere boundaries of two 3-discs (two solid balls).

It is simply amazing that anyone living in the 14th Century could contemplate a universe as a 3-sphere, let alone its slicing into 2-sphere slices. Think about it — more than six centuries passed before Einstein proposed that our universe (at any instant in time) may be viewed globally as a 3-sphere.

To supplement this chapter, which obviously contains no literary summary, let us sample the summary on page 485 of Volume 5 of the Macropædia in the 15th Edition of The New Encyclopædia Britannica. (Statements within square brackets "[·]" are mine.)

> **"The Divine Comedy."** ... A poet above all, he [Dante] felt that only in poetry would he be able to express fully his dream of a spiritual and civilized renewal of the whole of humanity. The poem, though unique, ... is inspired by the poetry of the Bible and by the Christian wisdom of the Holy Scriptures. Divided into three books, or *cantiche* (treating of Hell, Purgatory, and Paradise — ...) ... the number 3, a symbol of the Trinity, is always present in every part of the work, with its multiples and in its unity. ... The literal subject of the work is the journey Dante makes through the world beyond the grave, ... At age 35, on the evening of Good Friday, 1300, the poet finds himself wandering astray in a dark wood [allegorically very depressed]. After a night of anguish, he sets out toward a hill illuminated by the sun, but three wild beasts — a female ounce (a species of leopard), a lion, and a wolf: symbols of lust, pride, and avarice [greed for riches] — bar his path and force him back toward the darkness of wood. Virgil, however, sent by the Virgin Mary, St. Lucy, and Beatrice, appears to help him. He guides Dante through the infernal realm and the mountain of Purgatory, at the summit of which the Roman poet [Virgil] is replaced by Beatrice, who then conducts Dante (raising him from heaven to heaven by the brilliant and loving power of her glance, which is that of a blessed soul contemplating God) as far as the Empyrean [the abode of God], where the poet enjoys for a brief moment the supreme vision of the divinity.

For details surrounding Albert Ritter's and Gustave Doré's art works, Figures 2.2 and 2.4 respectively, visit Wikipedia on the Internet. (The computer files that generated these figures were obtained from Wikimedia Commons, a freely licensed media file repository.)

For additional references concerning Dante's 3-sphere see §A7 *Dante's 3-sphere construct.*

CHAPTER 3

Einstein and the 3-Sphere

In 1917 Einstein viewed the universe (at any instant in time) as a 3-sphere. This chapter concerns Einstein's comments surrounding the 3-sphere as expressed within his book *Ideas and Opinions*.[1]

His book was written for the general public, and Einstein expressed his opinion of such books in the following quote

> It is of great importance that the general public be given an opportunity to experience — consciously and intelligently — the efforts and results of scientific research.[2]

§14 IMAGINATION BOGGLES

In the quote below, Einstein acknowledges the challenge awaiting those who try to understand a 3-sphere.

> From the latest results of the theory of relativity it is probable that our three-dimensional space is also approximately spherical, that is, that the laws of disposition of rigid bodies in it are not given by Euclidean geometry, but approximately by spherical geometry, if only we consider parts of space which are sufficiently extended. Now this is the place where the reader's imagination boggles. "Nobody can imagine this thing," he cries indignantly. "It can be said, but cannot be thought. I can imagine a spherical surface well enough, but nothing analogous to it in three dimensions." We must try to surmount this barrier in the mind, and the patient reader will see that it is by no means a particularly difficult task.

---

[1] Einstein's book as published by Dell Publishing Co., Inc., ISBN: 0-440-34150-7, Fifth Laurel printing — June 1981, Copyright © MCMLIV by Crown Publishers, Inc. Author Albert Einstein, Based on MEIN WELTBILD, edited by Carl Seelig, and other sources with New translations and revisions by Sonja Bargmann. The quote begins on page 237 in the *Geometry and Experience* section of Part Five: *Contributions to Science*.

[2] The quote appears in the *Foreword by Albert Einstein* at the beginning of Lincoln Barnett's book *The Universe and Dr. Einstein*. Barnett's book is a Mentor Book published by The New American Library of World Literature, Inc. 501 Madison Avenue, New York 22, New York with Copyright 1948 by Harper & Brothers.

For background to the above quote, recall that the word *geometry* splits into *geo*, meaning "earth", and *metry*, implying "measurement." Einstein uses the phrase *Euclidean geometry* — the geometry that is the 3-dimensional extension of the *plane geometry* that we learned in secondary school, where we studied measurements relative to straight lines, where all triangles have the sum of their angles *equal to* 180°, etc.

And he references *spherical geometry* — the geometry of a 3-sphere which is the 3-dimensional analogue of *2-sphere geometry*, where "lines" are *great circles* — circles that contain diametrically opposite points, where all "triangles" have the sum of their angles *greater than* 180°, etc.

§15 Projecting S^1

Following his statements concerning the difficulty of "seeing" a 3-sphere, Einstein begins his quest to explain spherical geometry by considering the *projection of a 2-sphere*.[3]

Here, however, for background we shall at first step down one dimension and consider the *projection of a 1-sphere*.

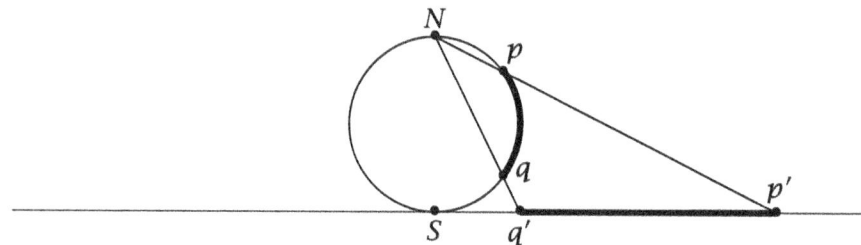

FIG. 3.1 Points p and q projected from S^1 into a line.

Think of the point N in Figure 3.1 as the *north pole*, which is diametrically opposite the *south pole S*, the point where our circle touches our line.

Except for the north pole N, we *project* each point on the circle into the line: Use a laser beam issuing from the point N and aimed at a point p on the circle. As illustrated, the beam starts at N and passes through the point p on the circle on its way to the point p' on the horizontal line. This is (*stereographic*) *projection*. The point p is projected onto the point p'. In reverse, p' is the *image* of p. Einstein calls p' the *shadow* of p. In this chapter we shall use "image" and "shadow" interchangeably.

In Figure 3.1 we also see q projected onto q', and it follows that the *curved segment* with endpoints p and q produces a corresponding *straight-segment* shadow whose endpoints are p' and q'.

---

[3]To explain the projection of a 2-sphere into a plane Einstein uses the picture on page 283 of his book.

Note that the length of the curved segment is only about half the length of its shadow. This observation shows that (stereographic) projection does not "preserve lengths" of segments as they are transformed (moved) from the circle into the line.

Turning to Figure 3.2, we see nine equal-length curved segments and their shadows. Observe that as the distances between the *equal-length* curved segments and the south pole increase, the lengths of the shadows increase.

In other words, in the context of line geometry (Euclidean geometry), as the distances between the shadows and the south pole increase, the lengths of the shadows increase. And as a fixed-length curved segment moves toward the north pole, then in the context of the line geometry the length of its shadow would approach infinity.

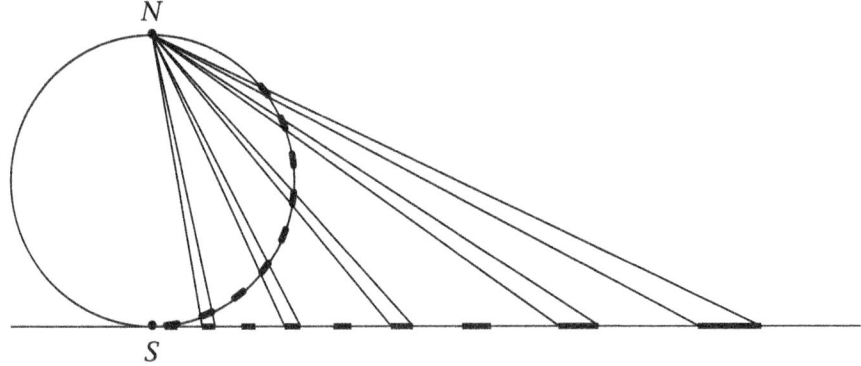

FIG. 3.2 Projection stretches equal-length curved segments.

An intelligent being living on the one-sphere (circle) would measure the curved segments using spherical geometry and find that all segments have the same length. Now let us suppose that the horizontal line is also a universe with an *unusual geometry* — a straight-line universe where measuring rods stretch as the rods are moved away from the point S. Furthermore suppose that their measuring rods stretch so that upon measuring "the shadows of the curved segments" they find that all of these "shadows" are also of the same length.

In such a case, the inhabitants of the straight-line universe with *unusual geometry* would be living in a world with *spherical geometry*. Stepping up by one dimension, Einstein uses a 2-sphere and shadows in a plane to explain how humans could realize spherical geometry; he then uses the spherical geometry of a 2-sphere to motivate how we may "think about" the geometry of a 3-sphere (inside of 4-dimensional space).[4]

As for a final observation, note that Figure 3.2 illustrates the general fact that length measurements in the circle universe (with spherical geometry) agree with length measurements in the line universe (with Euclidean geometry) when measuring *near* the south pole. For the curved segments in Figure 3.2 in particular, we see that the lengths of the two curved segments closest to S closely match the lengths of their shadows.

§16 Projecting S^2

Stepping up one dimension, we illustrate projection from a 2-sphere into a plane (Figures 3.3, 3.4, 3.5, and 3.6).

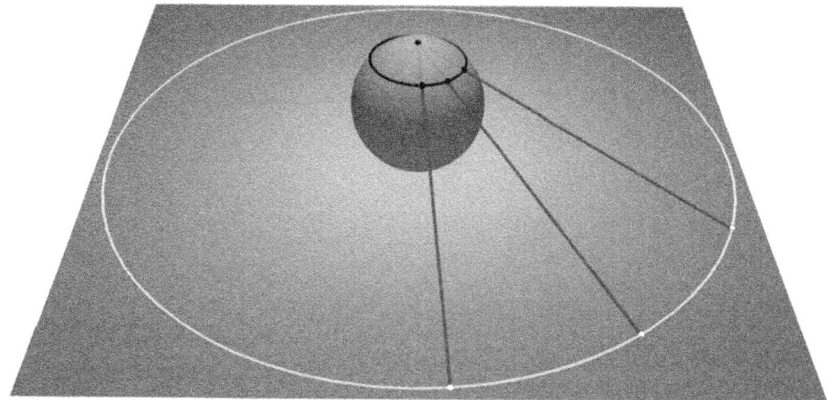

FIG. 3.3 Projecting a circle of constant latitude into a plane.

Again, the *north pole* — the black dot at the top of the sphere — is the only point that has no shadow. In Figure 3.3 we see a black circle of "constant latitude" whose shadow is the white circle. Note that the black circle in Figure 3.3 is not a *great circle* — a great circle on a 2-sphere has the same radius as the 2-sphere.

---

[4]Einstein also uses the words "Euclidean" and "non-Euclidean" to convey a distinction between geometries — Euclidean geometry is the geometry of human experience, i.e., the geometry of a line, a plane, or our visual 3-space that includes our usual "measurements." Euclidean geometry is the geometry that Euclid documented circa 300 BC. In contradistinction, spherical geometry differs from Euclidean geometry as illustrated above: If you lived on the line, then to experience spherical geometry on the line both you and your "measuring rod" would expand as you moved away from the south pole. And you would not realize the expansion because you could not measure it!

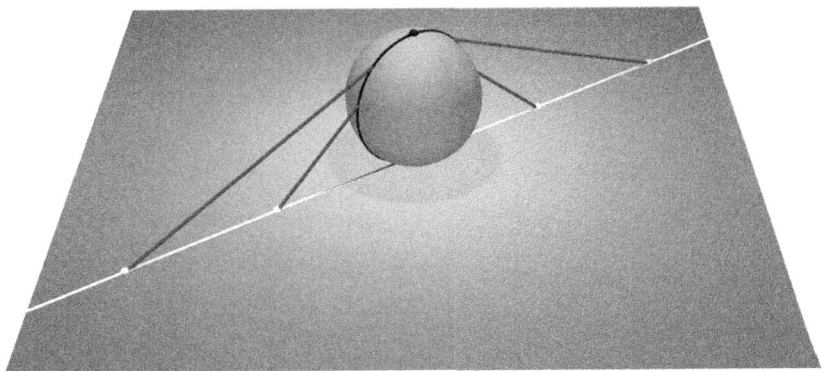

FIG. 3.4 Projecting a great circle of constant longitude into a plane.

And in Figure 3.4 we see the white-line shadow of a black *circle of constant longitude* — a constant-longitude circle is a great circle that goes through the north and south poles. A circle of constant longitude is sometimes called a *meridian*.

In order to continue the §14 Einstein quote, we need some notation and a graphic similar to the graphic on page 238 of Einstein's book. The similar graphic that contains the relevant notation is our Figure 3.5 where we see a curved 2-disc L bounded by the black circle on a 2-sphere K as well as the shadow L' of L.

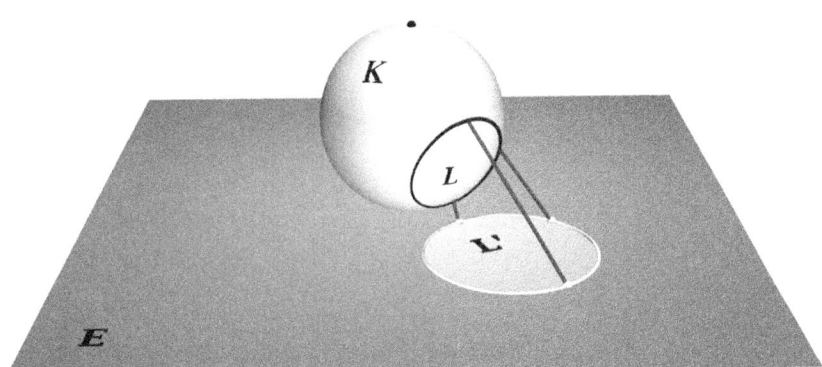

FIG. 3.5 A sphere K, a plane E, and a curved 2-disc L with shadow L'.

Continuing with the illustrated "curved 2-disc" L on the sphere K and its shadow L', we can also illustrate a feature that Einstein describes. In Figure 3.6 we see how the shadows of *same-sized* curved 2-discs expand as they move away from the south pole. Keep in mind that Einstein also uses "N" and "S" to denote the north and south poles of the 2-sphere K.

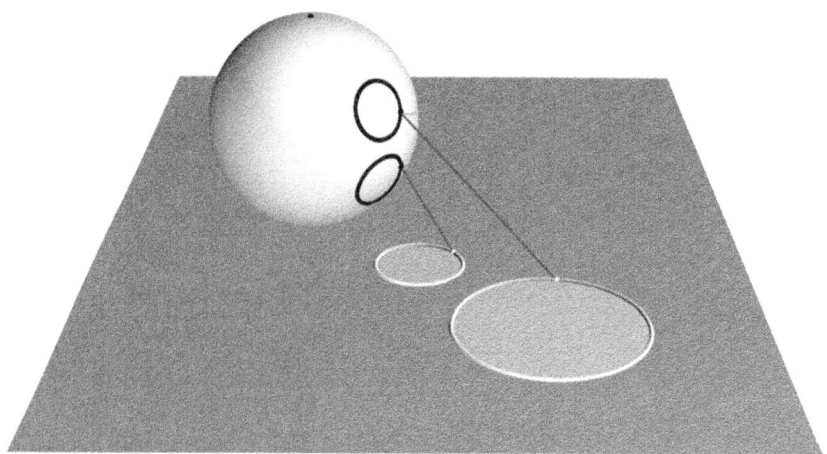

FIG. 3.6 Shadows of same-size spherical discs increasing in size.

With the background now in place, we continue the §14 Einstein quote:

> We must try to surmount this barrier in the mind, and the patient reader will see that it is by no means a particularly difficult task. For this purpose we will first give our attention once more to the geometry of two-dimensional spherical surfaces. In the adjoining figure let K be the spherical surface, touched at S by the plane, E, which, for facility of presentation, is shown in the drawing as a bounded surface. Let L be a disc on the spherical surface. Now let us imagine that at the point N of the spherical surface, diametrically opposite to S, there is a luminous point, throwing a shadow L' of the disc L upon the plane E. Every point on the sphere has its shadow on the plane. If the disc on the sphere K is moved, its shadow L' on the plane E also moves. When the disc L is at S, it almost exactly coincides with its shadow. If it moves on the spherical surface away from S upwards, the disc shadow L' on the plane also moves away from S on the plane outwards, growing bigger and bigger. As the disc L approaches the luminous point N, the shadow moves off to infinity, and becomes infinitely great.
>
> Now we put the question: what are the laws of disposition of the disc-shadows L' on the plane E? Evidently they are exactly the same as the laws of disposition of the discs L on the spherical surface. For to each original figure on K there is a corresponding shadow figure on E. If two discs on K are touching, their shadows on E also touch. The shadow-geometry on the plane agrees with the disc-geometry on the sphere. If we call the disc-shadows rigid figures, then spherical geometry holds good on the plane E with respect to these rigid figures. In particular, the plane is finite with respect to the disc-shadows, since only a finite number of shadows can find room on the plane.

§16 PROJECTING S^2 31

> At this point somebody will say, "That is nonsense. The disc-shadows are *not* rigid figures. We have only to move a two-foot rule about on the plane E to convince ourselves that the shadows constantly increase in size as they move away from S on the plane toward infinity." But what if the two-foot rule were to behave on the plane E in the same way as the disc-shadows L'? It would then be impossible to show that the shadows increase in size as they move away from S; such an assertion would then no longer have any meaning whatever. In fact the only objective assertion that can be made about the disc-shadows is just this, that they are related in exactly the same way as are the rigid discs on the spherical surface in the sense of Euclidean geometry.
>
> We must carefully bear in mind that our statement as to the growth of the disc-shadows, as they move away from S toward infinity, has in itself no objective meaning, as long as we are unable to compare the disc-shadows with Euclidean rigid bodies which can be moved about on the plane E. In respect of the laws of disposition of the shadows L', the point S has no special privileges on the plane any more than on the spherical surface.
>
> The representation given above of spherical geometry on the plane is important for us, because it readily allows itself to be transferred to the three-dimensional case.

So, using his projection of his 2-sphere K into his plane E (Figure 3.5), Einstein moves the "geometry" of the sphere (non-Euclidean) into a plane, preparing the reader for his argument of how to think about the "geometry" of the 3-sphere. Remember that by "Euclidean geometry" Einstein means the ordinary geometry where *parallel lines do not meet*, etc.

So to continue with the Einstein quote, we pause to note that he now steps up one dimension. That is, he uses "S" to denote "a point in ordinary 3-dimensional human visual space," and "sphere" to mean a *3-disc* (solid ball with a 2-sphere boundary). And L' denotes the 3-disc *shadow* of a 3-disc L. So parallel to the lower-dimensional cases, we now have a *shadow L'* that is the stereographic image of a solid ball L that lives inside a 3-sphere. Hence like the 2-dimensional case, L' is not rigid, but can change shape when L moves away from the point S on the 3-sphere. With this change in the meaning of the notation, let us return to Einstein's quote:

> Let us imagine a point S of our space, and a great number of small spheres, L' which can all be brought to coincide with one another. But these spheres are not to be rigid in the sense of Euclidean geometry; their radius is to increase (in the sense of Euclidean geometry) when they are moved away from S toward infinity; it is to increase according to the same law as the radii of the disc-shadows L' on the plane.

After having gained a vivid mental image of the geometrical behavior of our L' spheres, let us assume that in our space there are no rigid bodies at all in the sense of Euclidean geometry, but only bodies having the behavior of our L' spheres. Then we shall have a clear picture of three-dimensional spherical space, or, rather of three-dimensional spherical geometry.

§17 Einstein's view

Let us again pause from Einstein's quotes to point out that his last sentence quoted above, *Then we shall have a clear picture of three-dimensional spherical space, or, rather of three-dimensional spherical geometry*, indicates that he has provided "a view" of the geometry of a 3-sphere. So let us continue with his quote:

> Here our spheres must be called "rigid" spheres. Their increase in size as they depart from S is not to be detected by measuring with measuring-rods, any more than in the case of the disc-shadows on E, because the standards of measurement behave in the same way as the spheres. Space is homogeneous, that is to say, the same spherical configurations are possible in the neighborhood of every point.[5] Our space is finite, because, in consequence of the "growth" of the spheres, only a finite number of them can find room in space.
>
> In this way, by using as a crutch the practice in thinking and visualization which Euclidean geometry gives us, we have acquired a mental picture of spherical geometry. We may without difficulty impart more depth and vigor to these ideas by carrying out special imaginary constructions. ... My only aim today has been to show that the human faculty of visualization is by no means bound to capitulate to non-Euclidean geometry.

And with that last statement, Einstein ends his *Geometry and Experience* section in his book *Ideas and Opinions*.

§18 Comments

To be sure, the goal of this chapter was certainly not that of attempting to detail a part of Einstein's theory of relativity. Rather the goal here is simply to demonstrate, in Einstein's own words, that the 3-sphere played a role in science, just as the goal of the previous chapter was to demonstrate, using Dante's *Divine Comedy*, that the 3-sphere played a role in world literature and Christian faith.

---

[5] Einstein has a footnote at this point in his book. It reads as follows: This is intelligible without calculation — but only for the two-dimensional case — if we revert once more to the case of the disc on the surface of the sphere.

And these examples, however, are not exhaustive. Within mathematics the 3-sphere has a rich history. The reader who desires a challenge may consider the following article by R. H. Bing (1914–1986)[6]

> *Models for S^3*, The Collected Papers of R. H. Bing, American Mathematical Society, Providence, RI, 853–869.

Knowledge of the 3-sphere and at least part of its role in human history is a prerequisite for interest and appreciation of the "paint on canvas" *God's Image? (A partial image of a 3-sphere)*.

Finally, the pencil sketch of Einstein on page 24 was drawn by Darrin Lipscomb. The original sketch is approximately 30 years old and, prior to this publication, has never been copied.

---

[6]Bing was a "President of the American Mathematical Society" and a world-class research mathematician who wrote extensively about 3-dimensional manifolds and the 3-sphere.

CHAPTER 4

Einstein's Universe

It was 1917 when Albert Einstein proposed his model of the universe. In this chapter we discuss the definition of Einstein's Universe (EU) as presented in the article *Einstein's Static Universe: An Idea Whose Time Has Come Back?*[1]

The title tells all — EU was at first accepted, then questioned, and currently is *an idea whose time may have come back.* We are interested in Einstein's Universe because it may be expressed by saying, *at any instant in time our universe may be viewed as a* 3-*sphere.*

§19 Hollow Pipe

The article begins with a definition of Einstein's Universe (EU). To motivate the definition, however, we replace Einstein's 3-sphere S^3 with a one-sphere S^1, and then view the basic construct as a *hollow pipe*.

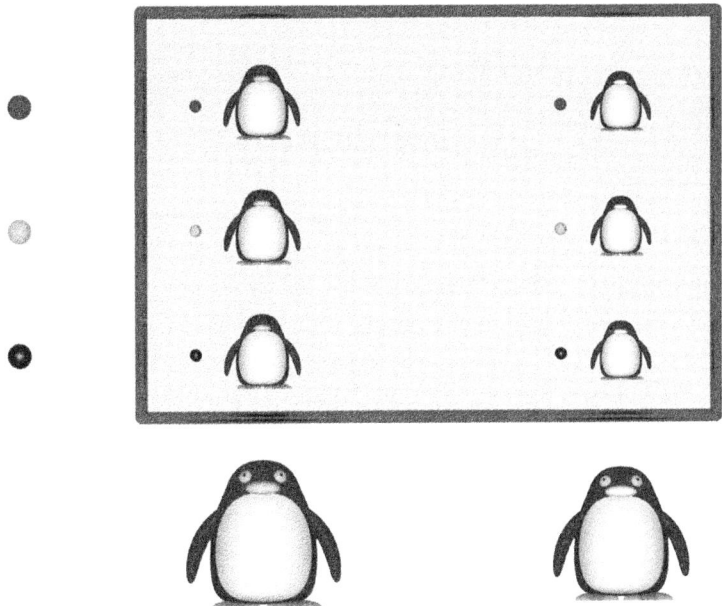

FIG. 4.1 Elementary-school example of a "Cartesian product" "$3 \times 2 = 6$".

---

[1] The article (subtitle *A tribute to Irving Ezra Segal* (1918–1998)) is authored by Aubert Daigneault and Arturo Sangalli. See Jan. 2001 American Mathematical Society *Notices*, page 9.

We start with the term *Cartesian product*.[2] With three marbles (dark-gray, light-gray, and black) and two penguins (Freddy and his smaller sister), we frame the six possible *pairs*. And each pair, say marble m and penguin p, may be pictured as a single entity, "point (m,p)", in the Cartesian product

$$\text{marbles} \times \text{penguins}.$$

This may sound like double speak, but the notation "(m,p)" as a single entity nicely names the pair m and p, and conversely, the pair m and p nicely conveys the single entity "(m,p)".

Our current concern is more general than *marbles* × *penguins*. We desire to picture $R \times S^1$ where we substitute the *Real timeline R* for the marbles, and the circle S^1 for the penguins. We are interested in creating a picture of all "points" in $R \times S^1$. In particular, while the three marbles are aligned in a straight line, the Real timeline is a real line.

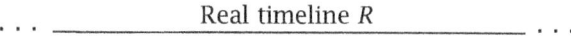

And just as the letter "m" denotes a marble, the letter "t" will denote *an instant of time*.

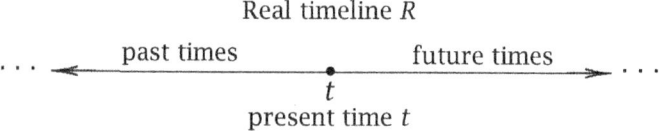

Moreover, for each marble m Figure 4.1 provides a corresponding *horizontal* row "$m\times$ *penguins*" that shows the "universe of penguins" paired with m. Analogously, with each instant t of time we desire a picture of the "row" $t \times S^1$ that contains the "universe of points" S^1 paired with t. Simply put, we desire a picture of $R \times S^1$ that shows each $t \times S^1$ (Figure 4.2).

It turns out that the entire Cartesian Product $R \times S^1$ with each of its "rows" $t \times S^1$ may be pictured as a *hollow pipe*.

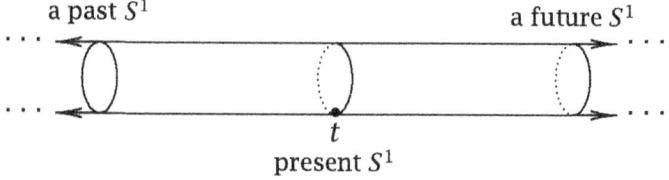

FIG. 4.2 The Cartesian product $R \times S^1$ as a hollow pipe of infinite length.

---

[2]*Cartesian* refers to René Descartes (1596–1650 AD) who "pictured" the "points in a plane" as "pairs of numbers" — the key word being *pairs*.

The hollow-pipe picture of $R \times S^1$ is a *faithful picture* because each point in the picture corresponds to one, and only one, point in $R \times S^1$. And we clearly see that distinct times t and t' induce distinct non-touching universes $t \times S^1$ and $t' \times S^1$.[3]

Now let us consider the dimension aspect when trying to see a hollow pipe. Suppose Freddy has either one- or two-dimensional vision, and his friend "Albert Pinstein" tells him about a hollow-pipe universe $R \times S^1$. It does not take much imagination to realize that Freddy cannot see $R \times S^1$.

FIG. 4.3 With 1- or 2-dimensional vision Freddy cannot see $R \times S^1$.

Moving up one dimension, we may similarly expect that humans with 3-dimensional vision cannot see $R \times S^2$.

§20 $R \times S^2$ AS A SOLID PIPE?

Just as we can picture $R \times S^1$ as a hollow pipe, we might try to step up one dimension and picture $R \times S^2$ as a *solid* pipe.

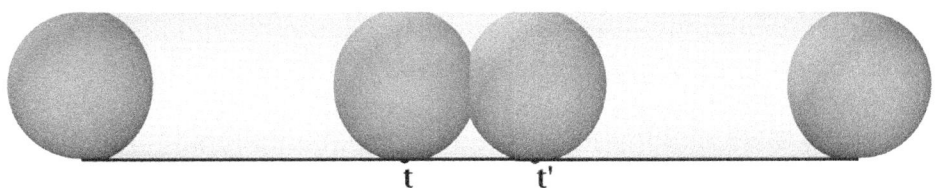

In other words, think of a bowling ball rolling down the gutter at a bowling alley. But a solid pipe is not a picture of $R \times S^2$ because, as we see below,

---

[3]In the context of Figure 4.1, this amounts to saying distinct marbles induce "distinct horizontal rows" — distinct rows have no "pairs" (points) in common.

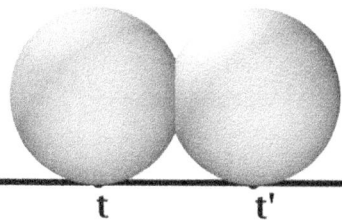

distinct instants t and t' of time do not induce non-touching $t \times S^2$ and $t' \times S^2$. So our plan to use a *solid pipe of infinite length* to produce a faithful picture of $R \times S^2$ fails.

To learn to think about $R \times S^2$ — *an organized line R of 2-spheres* — look back at Figure 4.3. On the left side we see Freddy with 2-dimensional vision looking at $R \times S^1$. In Freddy's case think about the line R, *which runs outside (vertically) of Freddy's visual plane while touching Freddy's visual plane only at a single point.* The line R adds an extra dimension to Freddy's visual plane.

Analogously, consider $R \times S^2$: We step up one dimension, seeing S^2 with our 3-dimensional vision. But to see $R \times S^2$ we, like Freddy, need a line R that runs outside of our visual plane while touching our visual plane only at a single point. That is, we need an extra dimension to see $R \times S^2$.

§21 $R \times S^3$

Turning to $R \times S^3$, we cannot visualize $R \times S^3$ because we cannot visualize S^3. Nevertheless, with the $R \times S^1$ hollow pipe picture coupled with our understanding of the need for an extra dimension to see $R \times S^2$, we have enough background to understand EU $R \times S^3$. Simply think of EU as a Real timeline R that has a uniform placing of a 3-sphere S^3 at each instant t of time. That is, the Cartesian product $R \times S^3$ is the collection $t \times S^3$ where t takes on all values of time — past, present, and future.

So we arrive at a quote from the Einstein's Static Universe article. Note that since the authors refer to "physical space" as "finite and curved" it is clear that they are speaking of the 3-sphere S^3 in the Cartesian product $R \times S^3$ because the R factor is infinite.

> ... In 1917 Albert Einstein proposed a model for space-time known as the Einstein Universe (EU), in which the totality of physical space is finite and curved.[4] "Nothing in general relativity has intrigued the lay public more than Einstein's possibility of a closed, finite spatial universe," observed Theodore Frankel in his 1979 introductory book to Einstein's theory.

---

[4]"Cosmological considerations on the general relativity theory", pp. 177-188 in the *The Principle of Relativity*, by H.A. Lorentz, A. Einstein, H. Minkowski, and H. Weyl, with notes by A. Sommerfeld, published unaltered and unabridged by Dover Publications, Inc. in 1952.

In EU, space may be mathematically described as a three-sphere S^3 of fixed radius r, i.e., as the boundary of a four-dimensional ball, by the equation $u_1^2+u_2^2+u_3^2+u_4^2 = r^2$. In Einstein's model, time had no beginning: it is infinite in both directions, and so the universe has always existed and will always exist. Hence EU may be presented as the Cartesian product $R \times S^3$ where R is the whole real timeline. ...

§22 COMMENTS

Again, the goal here is not to explain modern cosmology. Rather, the goal in this chapter is the same as it has been in the previous chapters — *show relevance of the 3-sphere.*[5]

---

[5] Part 2 of Appendix 2 provides a continuation of our cursory overview of *Einstein's Static Universe: An Idea Whose Time Has Come Back*. Evidently the theory of *chronometric cosmology* says that Einstein's Universe may indeed be the correct model of our universe.

CHAPTER 5

Images of S^1 and S^2

Human experience allows us to comfortably view photographs of familiar objects. A camera can capture an image of a puppy, and because of our experience with puppies we can comfortably — without pause of having to think about what we are seeing — enjoy and understand the picture.

On the other hand, suppose we "see" a partial picture of a 3-sphere. *How could we judge it?* We may try to judge it as an extension of S^1 and S^2. But *how does anyone judge something that has never been seen?*

In this chapter we use S^1 and S^2 to set the stage for our partial picture of a 3-sphere.

§23 PICTURING A ONE-SPHERE

Do you recall our elementary-school days when those among us who were not artistic learned to *connect the dots*? We could use *graph paper* — horizontal and vertical grid lines — to capture a picture as a collection of dots (Figure 5.1); and then we revealed the picture by "connecting the dots."

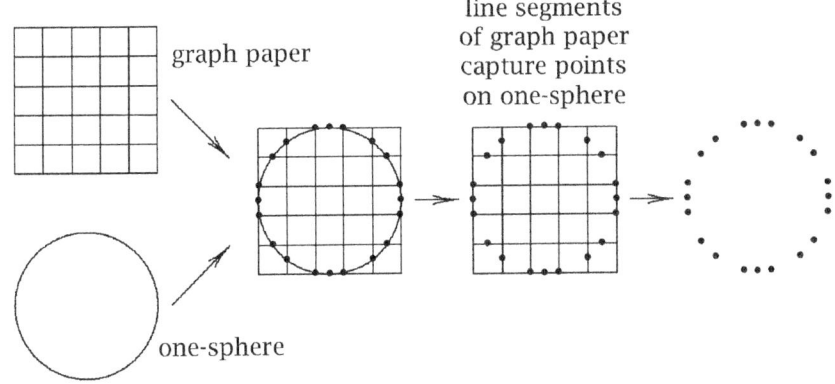

FIG. 5.1 Using graph paper to picture a one-sphere (circle).

At the risk of oversimplification, the graphic above provides a bird's-eye-view of our approach to creating a partial picture of a 3-sphere. The 3-sphere also requires "graph paper," but the graph paper must be designed for 4-space (4-dimensional space).[1]

One feature of any graph paper is that of *distortion*, e.g., the grid in Figure 5.1 may be distorted (Figure 5.2), which induces a movement of the *dots*.

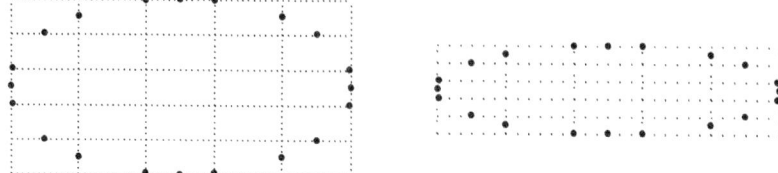

FIG. 5.2 Distorting the graph paper distorts the "cells" and moves "the dots".

The distortions pictured above are *faithful representations* because each distinct pair of dots in Figure 5.1 are moved to correspondingly distinct dots in Figure 5.2. *Faithful* and *non-faithful* representations are illustrated below where, starting with distinct points p and q on a one-sphere, we deform the one-sphere until the points p and q are not distinct, which yields a non-faithful representation of a one-sphere (Figure 5.3).

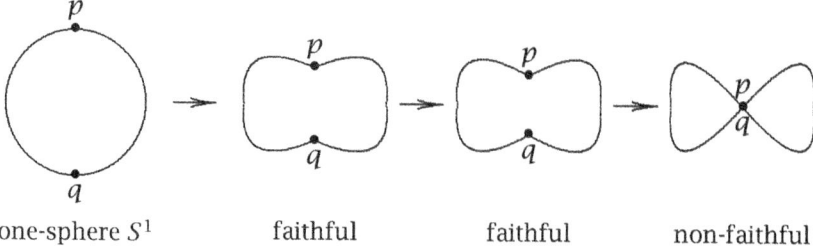

one-sphere S^1 faithful   faithful   non-faithful

FIG. 5.3 Two faithful and one non-faithful representations of a circle.

As for *an arrangement of dots* that partially picture S^3, let us return to our "*How could we judge it?*" question. Well, we can judge any arrangement of dots as to whether the representation is *faithful* or *non-faithful*. It turns out that the *collection of dots* that form our partial picture of S^3 is indeed a faithful representation — we know that *each dot within the picture represents one and only one point on the 3-sphere*.

Another feature of any such picture is the number of *dots*, which is determined by the density of the small squares (*cells*) that define the graph paper. Generally more cells yield more dots. To increase the number of cells we *shrink each cell by 1/2 toward each of its four corners* and then *replace the original cell with the four smaller cells*:

---

[1] Such graph paper is the topic of the following chapter. Our *4-space graph paper* is based on the *4-web*, a structure that lived only in 4-space until 2003. Think of a spider who lives in 4-space weaving its web according to the 4-web design.

For an example look at the 25-cell grid (5 × 5 squares) in Figure 5.1. By subdividing each of the 25 cells, we increase the grid to 100 = 25 × 4 cells, and then subdividing yet again, we increase the grid to 400 = 100 × 4 cells. These *rectangular-cell grids* are illustrated in Figure 5.4.

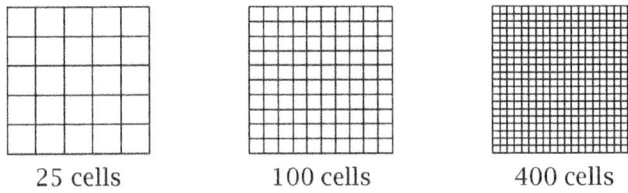

FIG. 5.4 Numbers of cells in sample rectangular-cell grids.

By increasing grid lines we may increase the number of dots captured. But *Could there be another type of graph paper that uses more than just vertical and horizontal grid lines?* The answer is yes.

§24 TWO-WEB GRAPH PAPER

Unlike the rectangular grid whose cells are squares, the cells of a *2-web* grid are triangles. But the rules for subdivision are similar. To increase the number of triangular cells, we (*a*) *shrink each cell by 1/2 toward its three corners*,

and (*b*) *remove the "middle triangle" from further subdivision*. The process is illustrated below. The first subdivision of a single cell contains three cells that subdivide and one cell (a hole) that does not subdivide. Notice that only the triangles that point toward the top of the page are further subdivided.

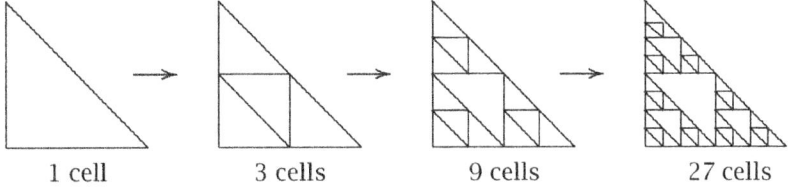

As for capturing points on S^1, we capture more points with more cells.

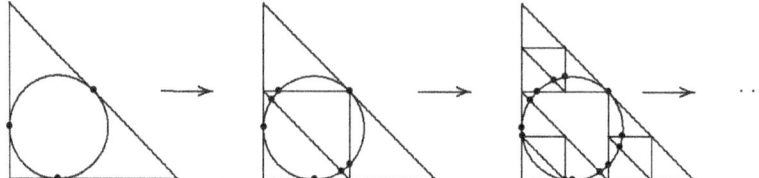

Two-web grids involve not only vertical and horizontal lines, but also slant lines. Each subdivision adds line segments that are parallel to one of the sides of the original triangle. So there are three groupings of lines — the vertical group, the horizontal group, and the slant group.

As illustrated below, each of the three groups of parallel lines slice an inscribed circle.

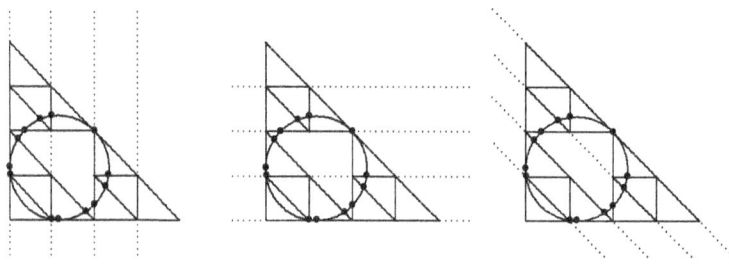

For artistic value we include below a 2-web grid with five subdivisions:

§25 Picturing a two-sphere

In order to capture points on a 2-sphere we start by placing the sphere inside of a *cube* — an empty box with equal-length edges.

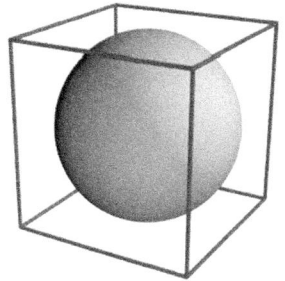 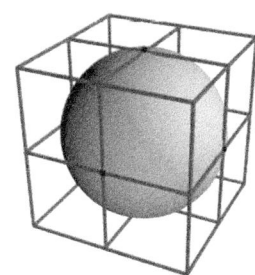

The graphic above illustrates the *original cell* (cube) and *its subdivision* comprised of 8 smaller (cubic) cells. Again, the subdivision involves shrinking the original cell toward each corner, i.e., we simply *shrink the original cube by 1/2 toward each of its eight corners*.

On the grid of our subdivision, the *cubical 8-cell grid*, we see three rather small black dots. These black dots mark the points (within our vision) where our 8-cell grid touches the sphere.

To capture more points on the sphere, we need more cells, hence more grid lines. So we subdivide each of the eight cells in the 8-cell cube into eight smaller cells, thereby obtaining a cubical 64-cell grid. The 64-cell grid can be further subdivided ad infinitum. All such structures are referred to as *cubical grids*.

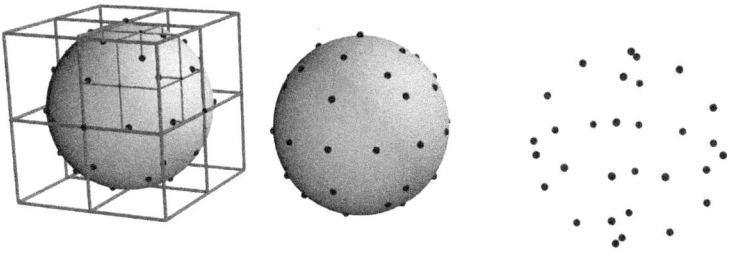

FIG. 5.5 Capturing points on a 2-sphere using a cubical grid.

For extra clarity, notice that among the eight relatively large cubical cells in Figure 5.5 only the top right-front cell has been subdivided. Had the other seven relatively large cells been likewise subdivided, there would be a total of 64 cells. The illustration shows where the grid lines of the 64-cell cube penetrate the 2-sphere. The dots alone are somewhat confusing because the picture does not provide enough information for the mind to process a sense of depth. But as one can see in Figure 5.6, the equatorial circle and the two meridian circles allow the mind's eye to organize the dots.

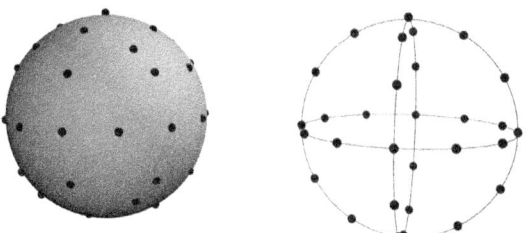

FIG. 5.6 Equatorial circle, and two meridian circles (circles through poles).

As with the rectangular grid, a distortion of the cubical grid produces a distortion of the faithful-dot representation (Figures 5.7 and 5.8).

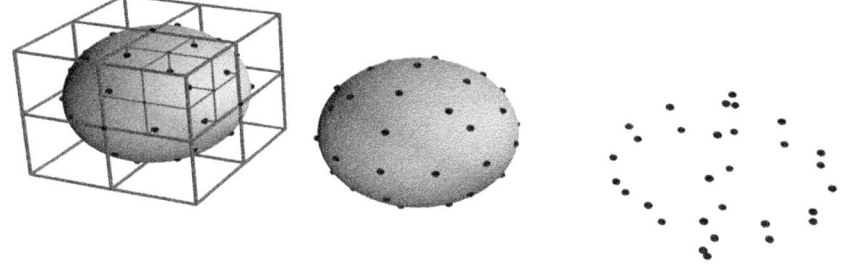

FIG. 5.7 When the grid stretches horizontally, the faithful-dots image follows.

Since the 2-sphere has no *prefered direction*, i.e., a 2-sphere can be rotated so that any point on its surface moves to the position of any other point on its surface, one can guess what happens when we stretch the grid vertically.

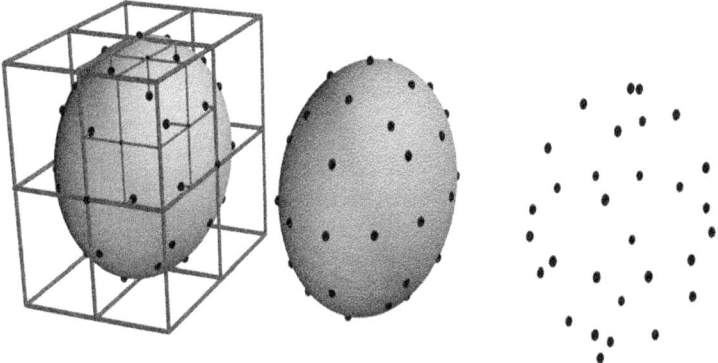

FIG. 5.8 When the grid stretches vertically, the faithful-dots image follows.

Notice that whenever we stretch the 2-sphere it is still recognizable. In essence, we view the stretching like we view human faces. You may have heard a hair dresser say, *You have a broad face and should have a close cut,* or, *You have a thin face and should fluff your hair.* Such statements provide support for the fact that humans view slight distortions as common.

§26 Three-web graph paper

A 3-*web* grid is based on the *tetrahedron*, pictured below as the *original cell*.

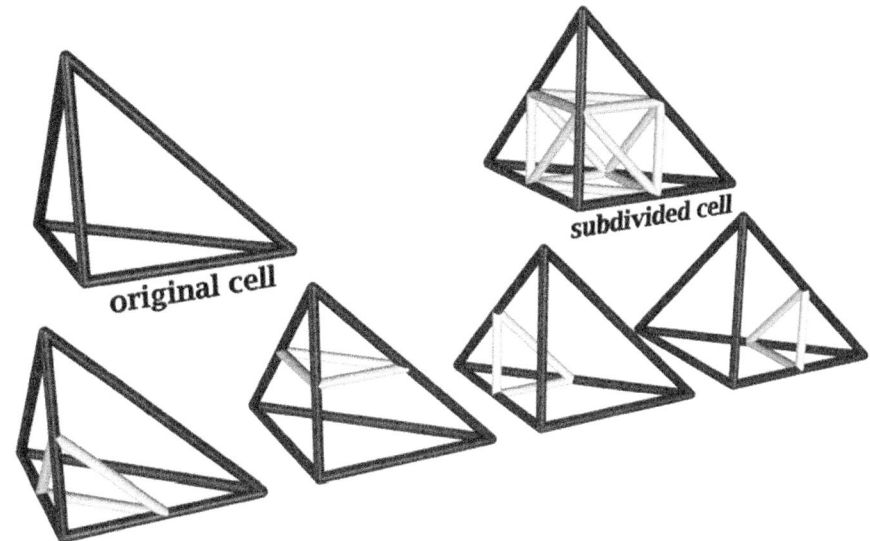

The word *tetrahedron* breaks into *tetra* meaning "four" and *hedron* meaning "face." Notice the *four triangular faces* — one horizontal, the two vertical faces that share the lone vertical edge, and one slant. The slant face is the face on the back side of the tetrahedron. The *subdivided cell* is obtained by *shrinking the original cell by 1/2 toward each of its four corners*. The four individual shrinkings are illustrated along the bottom of the graphic.

Like the grids discussed above, the 3-web grid is developed by an iteration of subdivisions. A tetrahedron, its first subdivision, and its second subdivision are pictured below: In this case the slant face is the front face.

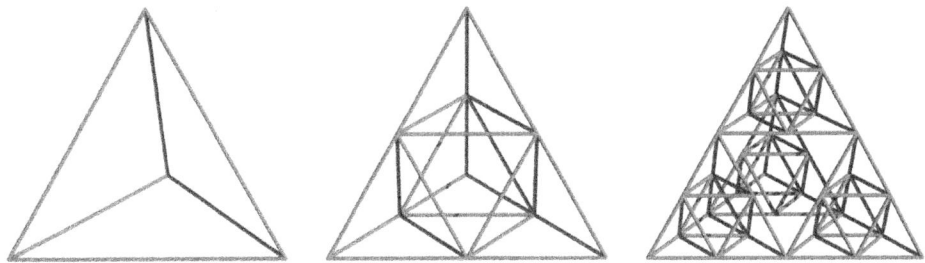

Turning to capturing dots on a two-sphere, we begin with Figure 5.9. The seven black dots are points where the lone tetrahedron touches the 2-sphere.

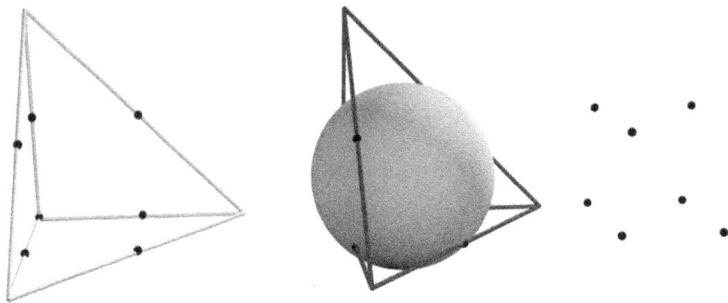

FIG. 5.9 Tetrahedral cell captures points on a 2-sphere.

On the surface, seven black dots do not convey much information. But consider the first subdivision of our tetrahedron. Using two parallel planes, we can organize six of the seven black dots (Figure 5.10).

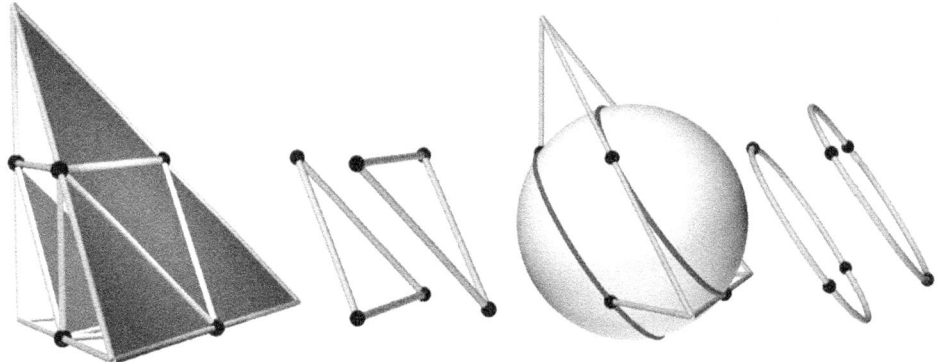

FIG. 5.10 The first subdivision of our tetrahedron provides slices of S^2.

And using the 3-web grid that is the second subdivision of our tetrahedron, we obtain more dots (Figure 5.11).

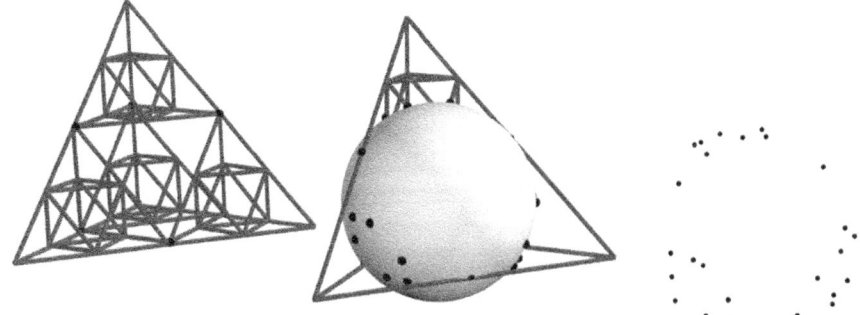

FIG. 5.11 The second subdivision of our tetrahedron captures more points.

Even if we continue to subdivide, thereby increasing the number of captured dots, we shall always have relatively large areas of S^2 that lie "outside" of our 3-web grid. So we ask, *Is there something about the captured dots that indicate that the span of the graph paper is insufficient?*

The answer, at least for the 2-sphere, is yes. Since the dots are points in our visual 3-dimensional space, we may "walk around the assemblage of 25 dots" looking for "flat areas" or "missing parts." The idea is illustrated in Figure 5.12.

FIG. 5.12 Circular, missing part, and flat-side views of 2-sphere dots.

Keep in mind that each of the three "assemblages of 25 dots" pictured above is an *image* of a 2-sphere. The 25 dots contain subtle information.

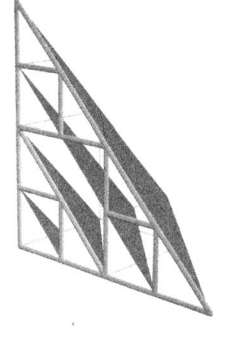

Since we know that the object in question is a 2-sphere, we know that the *slant planes* induced by the 3-web grid slice the sphere. When we subdivide the 3-web grid, we have many more planes. For example, notice the "slant" planes illustrated on the right. These planes are parallel. This means that we can slice the 2-sphere in parallel slices as illustrated in Figure 2.9. Similar observations follow for the two sets of vertical planes and the one set of horizontal planes.

In addition to subdividing to obtain more and more cells, we may move the 2-sphere, increase its size, decrease its size, or use "deformed" 3-web grids. For artistic value we include a 3-web grid with three subdivisions:

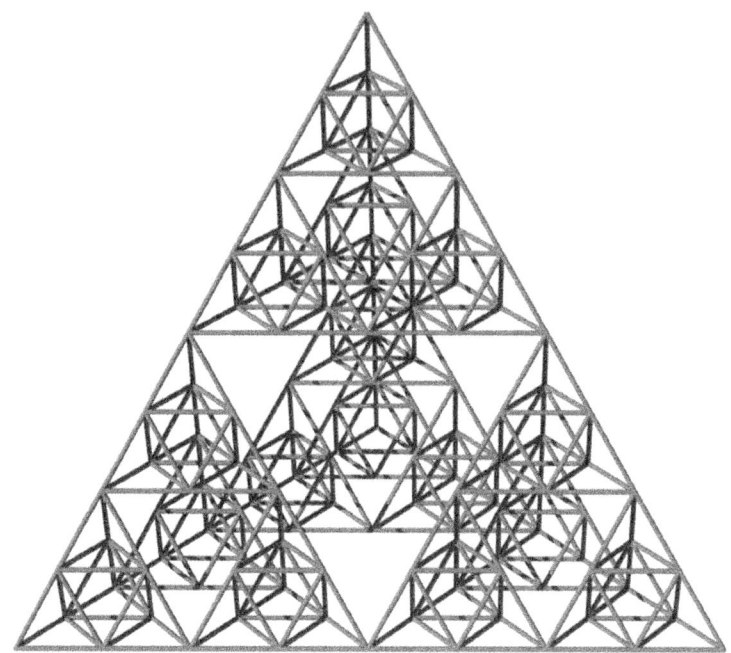

§27 Comments

The words 2-*web* and 3-*web* refer, respectively, to the *Sierpiński triangle* and the *Sierpiński cheese*. Sierpiński introduced his triangle circa 1900, and today it is arguably the most popular *classical fractal*. The triangle lives in a plane, and stepping up one dimension, the cheese is the extension of the triangle to 3-space (human visual space).

The graphic of the 2-web grid that appears at the end of §24 was generated with a formula developed by Chris Dupilka.

The following chapter concerns the 4-*web*, which extends the 3-web to the next higher dimension, namely 4-space. Originally, it was believed that the 4-web could not be faithfully moved into human vision. But in 2003, James Perry and Stephen Lipscomb did find a way to faithfully move the 4-web into human vision. The relevant article appeared in the Houston Journal of Mathematics, Volume 29, Number 3, on pages 691–710. The title of the article is *The generalization of Sierpiński's triangle that lives in 4-space*.

The illustrations of the 2-web grids in this chapter were restricted to a triangle (original 2-web cell) that had a 90^0 angle and two sides of equal length. The idea of an original 2-web cell, however, is not restricted to these triangles — one may start with any triangle. Similarly, while the illustrations of the 3-web grids in this chapter were restricted to a special tetrahedron, one may start with any tetrahedron.

To motivate the following chapter, look at the 2-*sphere picture of dots* that appears on the right side of Figure 5.5. If the picture of dots were presented totally isolated and without context, then we *might* be able to guess at what the dots represent.

In contrast, experience trying to guess what is conveyed in the illustration below. Look at the assemblage of dots and suppose you know that the picture of dots is faithfully represented.

What do you experience? That is the question! And after you have given some thought to your answer, *Is your mind's eye open to suggestion?* These are the kinds of questions that arise and initiate *converse art* — we begin, so to speak, with "paint on canvas" in the form of an "assemblage of dots" and then an observer must somehow organize the "assemblage of dots" in his mind's eye. When he is convinced of "what the assemblage of dots represents," he *decides on a name*.[2]

---

[2] The word *converse* means "in reverse order," and so the phrase *converse art* concisely conveys the idea that the "name" of the art is selected *after* one has the "paint on canvas".

CHAPTER 6

Four-web Graph Paper

We see what we see because of where we were when. As children our vision sends pictures to our brain, and our brain processes the information so that *pictures make sense*. We learn to put a square peg in a square hole, and then we move on.

Throughout human history, however, certain "thought problems" — problems that lie outside of common everyday experience — arise. The *fourth dimension* is a prime example.[1]

The fourth dimension is also basic to the concept of a 4-*web* which originally lived only in 4-dimensional space. Today, even non-mathematicians can see the 4-web structure and its grids, which is the focus of this chapter.

For completeness, we begin with the *hypercube*, which also lives in 4-dimensional space as the analogue of the 3-cube studied in §25.

§28 WHAT IS A HYPERCUBE?

Inside of 4-dimensional space, the basic cell is a *hypercube*, which is the 4-dimensional analogue of the 3-cube. To appreciate the graphic representation of the hypercube in Figure 6.1, let us start with a 0-cube, say the point p, and add dimensions until we arrive at the hypercube:

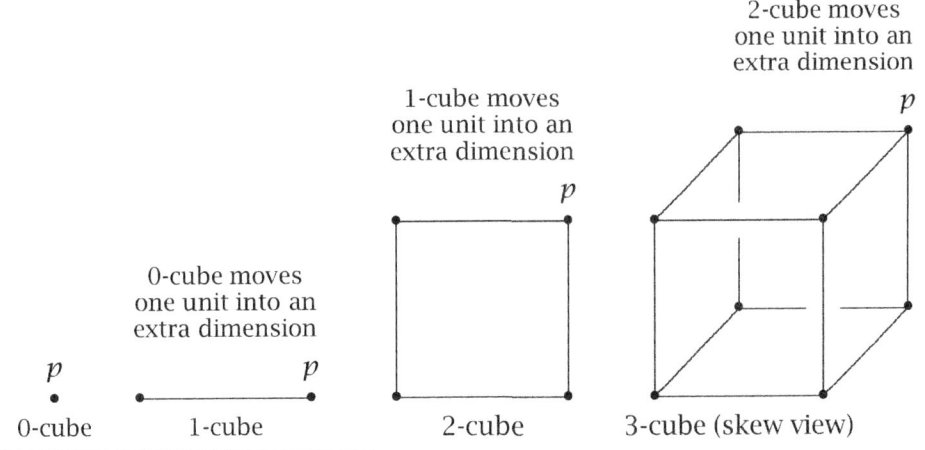

[1] "... the fourth dimension arose quite accidentally, from a failed attempt to create *three-dimensional numbers*" is a quote from John Stillwell's book cited in the first paragraph in §7. For more details see Stillwell's chapter titled *The Fourth Dimension*.

To construct a *one-cube* from the 0-cube, we push the point p one unit into an *extra dimension*. The resulting one-unit line segment together with its endpoints is a one-cube. To construct a *2-cube* from a 1-cube, we push the 1-cube (line segment) one unit into an extra dimension. The resulting square is a 2-cube. To construct a *3-cube* from a 2-cube, we push the 2-cube one unit into an extra dimension. The resulting *solid cube* — think of an ice cube — is a 3-cube.

To be sure we understand the process, let us take another look at the construction of the 3-cube from the 2-cube — notice in the graphic above that we see only a *skew view* of a 3-cube. Below, however, we provide a *face view* of the process of creating a 3-cube.

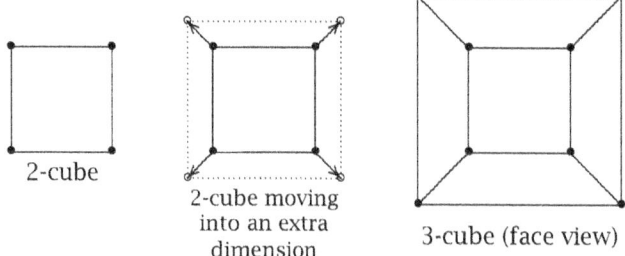

Turning to the hypercube, we note that pushing a 3-cube into an extra (fourth) dimension involves pushing each of its six faces (each face is a square) into an extra dimension. And it is the *face view* construction above that appears six times in our picture of the hypercube.

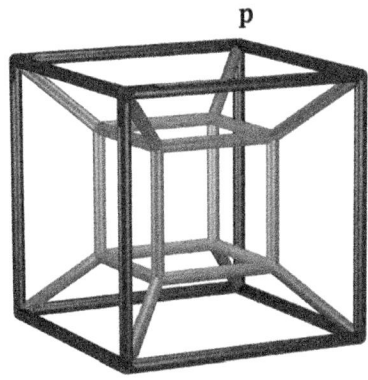

FIG. 6.1 A 3-dimensional representation of a hypercube (4-cube). By pushing the corners of the gray 3-cube toward the corresponding corners of the black 3-cube, we represent moving the gray 3-cube one unit into an extra dimension. Note that point p on the corner of the black cube relates to the first graphic in this chapter — the point p started (before the push into the extra dimension) on the corresponding corner of the gray cube.

This hypercube representation illustrates the fact that while we may experience only slight problems with cubical grids at low-dimensions — dimensions one, two, and three — we may experience significant conceptual problems with cubical grids at higher dimensions.

For example, the "extra dimension" used to construct a 4-cube creates 16 "corners" inside of 4-space. But its 3-space representation in Figure 6.1 contains only 8 (black) "corners".

§29 WHY IS IT THAT WE CANNOT SEE A HYPERCUBE?

To understand why human vision does not allow us to picture a *hypercube*, the analogue in 4-space of a 3-cube, let us consider the following 3-cube analogy: Below we see Freddy the penguin with his 2-dimensional vision.

FIG. 6.2 Freddy with 2-dimensional vision looks at a 3-cube.

Freddy can only see a cross section (a square) of the cube as pictured on the right. And to move off of his viewing plane, say in an "up-or-down direction" requires an extra "up-down dimension". Freddy's vision, unlike human vision does not include this extra dimension. So Freddy cannot see the cube, while humans easily see the cube. Just as part of the 3-cube exists outside of any 2-dimensional plane, the 4-cube exists outside of any 3-dimensional *hyper*plane (Figure 6.2).

Intuitively at least, this example with Freddy motivates the belief that humans (with 3-dimensional vision) cannot see the 4-cube, i.e., the hypercube.

§30 THE 4-WEB AND 4-WEB GRIDS

In the previous section we considered squares. In this section it is triangles. Inside of 4-dimensional space there exists a 4-web cell that is the analogue of the 3-web cell. Our approach runs parallel to our approach at picturing the hypercube. So recall §28, where we progressed with pictures of the 0-cube, 1-cube, 2-cube, and 3-cube. With a similar graphic below we progress with pictures of the basic cells for a 0-web, 1-web, 2-web, and 3-web.

A comparison of the pictures of the structures in these two lists shows that the latter list is easier to conceptualize. The reason for the simplicity is simply that instead of pushing a complete copy of a cube into an extra dimension, we merely place *one extra point in an extra dimension.*

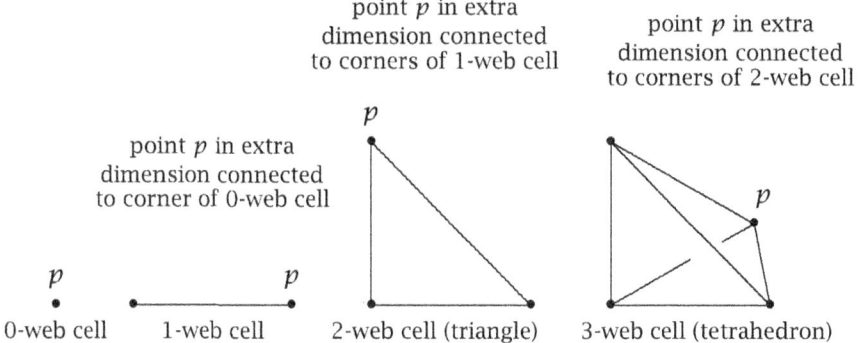

The 4-web cell is represented as the left-side graphic in Figure 6.3. The cell has five corners that partition into two groups — the *equatorials* and the *polars*. The *north pole* is the top corner, and the *south pole* the bottom.

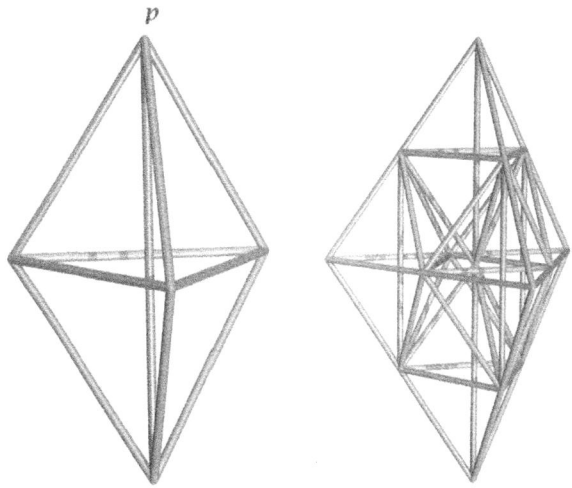

FIG. 6.3 The 4-web cell and its first subdivision.

The south pole and the three equatorials specify a 3-web cell (tetrahedron) within our 3-dimensional visual space, while the north pole represents the point p in the extra (fourth) dimension. Observe that, as required, the point p is connected via line segments to each corner of our 3-web cell. Also observe that the segment connecting the north and south poles passes through the *equatorial triangular hole*. The cell contains 10 line segments.

The first subdivision of our 4-web cell appears in Figure 6.3 (right-side graphic). It contains 50 line segments and is comprised of five smaller cells, each obtained by shrinking — *shrink the original cell by 1/2 toward each of its five corners.*

These five shrinkings appear in Figure 6.4: The first graphic on the left represents the shrinking toward the north pole, the adjacent graphic the shrinking toward the south pole, and the remaining graphics the shrinkings toward the equatorial corners.

When these small cells are combined to form the first subdivision with 50 segments, it is rather difficult to distinguish individual small cells.

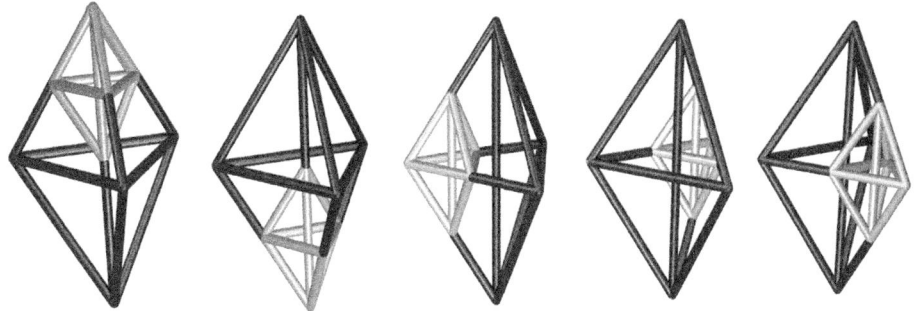

FIG. 6.4 The 4-web cell shrinks by 1/2 toward each of its five corners.

We can, however, distinguish individual small cells by changing the *line-segment* representation to one of *solids*. The change is detailed step-by-step within Figures 6.5 through 6.7.[2]

Each of the five small cells may be viewed as a *solid hexahedron* (a solid with six faces). The edges of a hexahedron correspond to nine of the 10 edges that define a 4-web cell — the line segment connecting the poles is inside of the solid hexahedron.

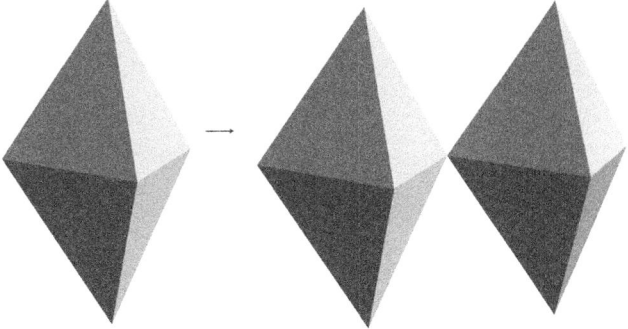

FIG. 6.5 One hexahedron, then two, each just touching the other.

In Figure 6.5 we see a lone small cell on the left, and on the right side we see two such cells *just touching* — *the two distinct cells have exactly one point in common.*

---

[2]Figures 6.5 through 6.10 are basically those that appear on pages 13 through 16 in the author's book *Fractals and Universal Spaces in Dimension Theory*, which was published in the "Springer Monographs in Mathematics" series in 2009 and reviewed in Edgar [13].

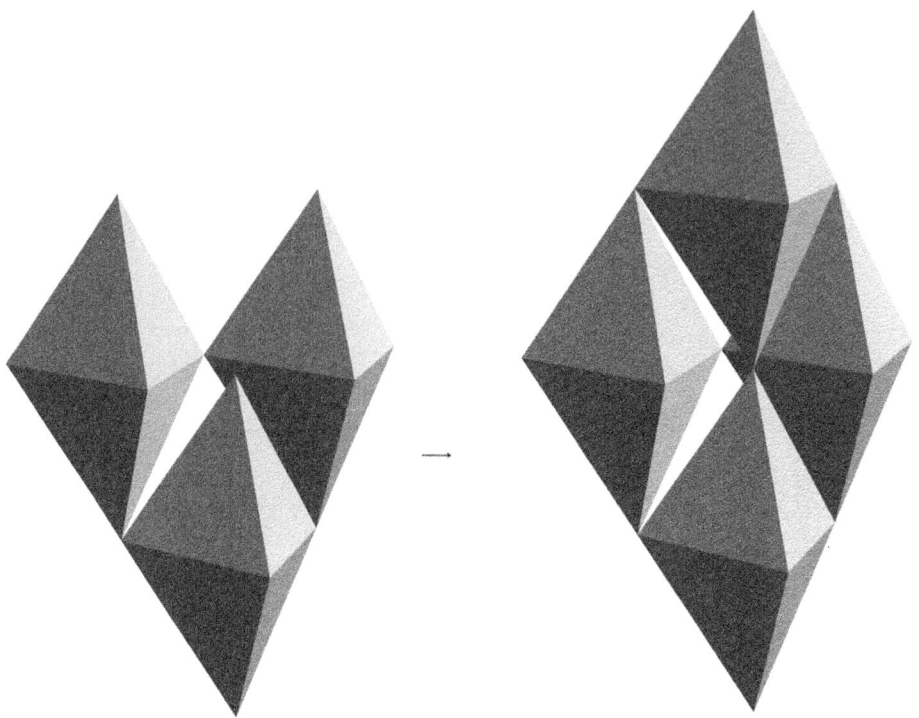

FIG. 6.6 Three, then four hexahedra, each just touching the others.

FIG. 6.7 Five hexahedra, each just touching the others.

§30   THE 4-WEB AND 4-WEB GRIDS   59

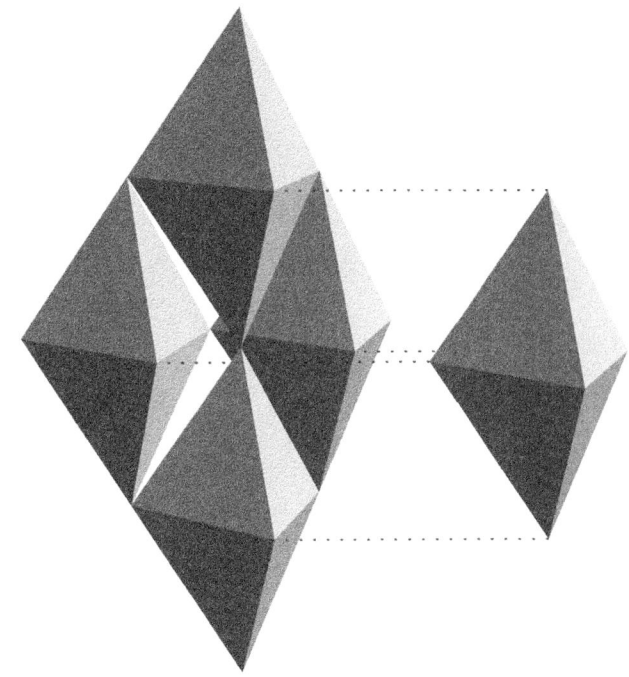

FIG. 6.8 Fitting the fifth hexahedron.

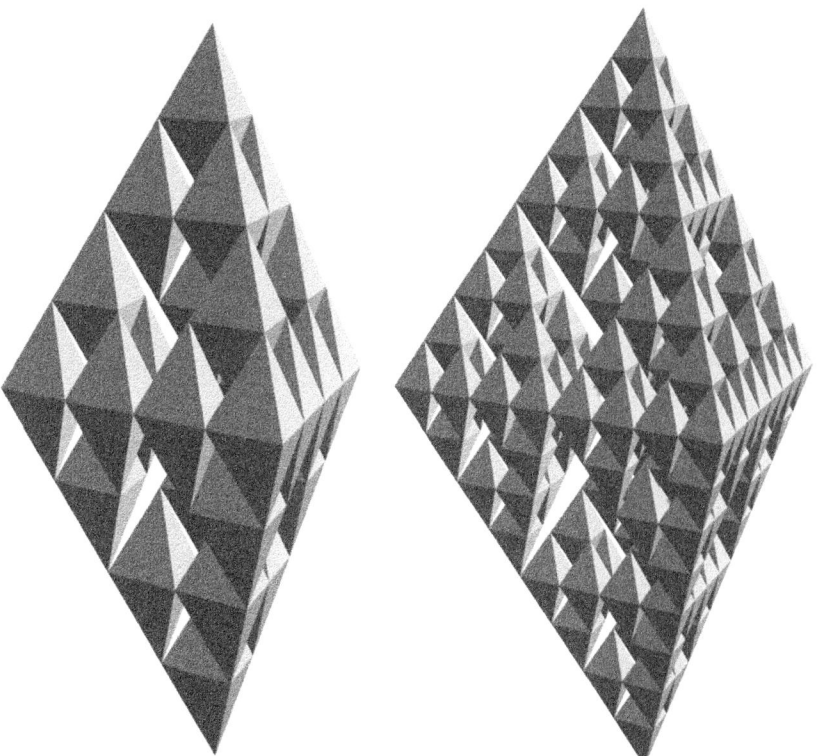

FIG. 6.9 Using solids to picture the second and third subdivisions of the 4-web cell.

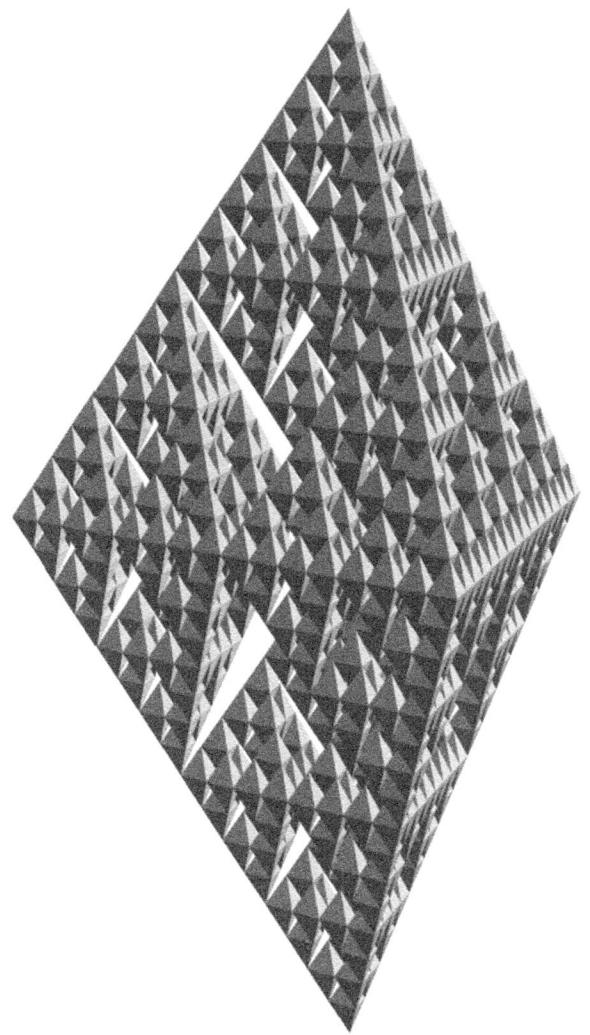

FIG. 6.10 Using solids to picture the fourth subdivision of the 4-web cell.

So in the previous few pages we have used solid hexahedra to illustrate approximations of the first, second, third, and fourth subdivisions of our 4-web cell. In turn, the subdivisions are crude approximations of the 4-*web*.

There is something of note here. Notice that each subdivision contains not only smaller hexahedra, but also more holes. In fact, as subdivisions continue the induced holes become dominant and the subdivisions "converge" to a *one*-dimensional object, namely, the 4-*web*.

To think about a 4-*web*, think of *a highly organized assemblage of line segments — each edge and polar segment in any hexahedron within any subdivision is a line segment of the 4-web*. In contrast, only those edges and polars of the hexahedra in a given subdivision form a 4-*web grid*. Such a grid appears in Figure 6.11.

FIG. 6.11 Chris Dupilka's graphic of a 4-web grid — 6th subdivision of a 4-web cell.

Dupilka's example of a 4-web grid (Figure 6.11) contains 156,250 line segments. To experience the density of these segments, let us move in toward the foremost equatorial corner (Figure 6.12).

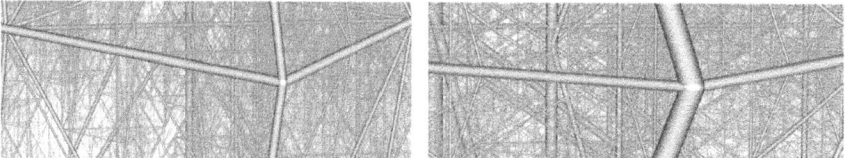

FIG. 6.12 Moving toward the foremost equatorial corner.

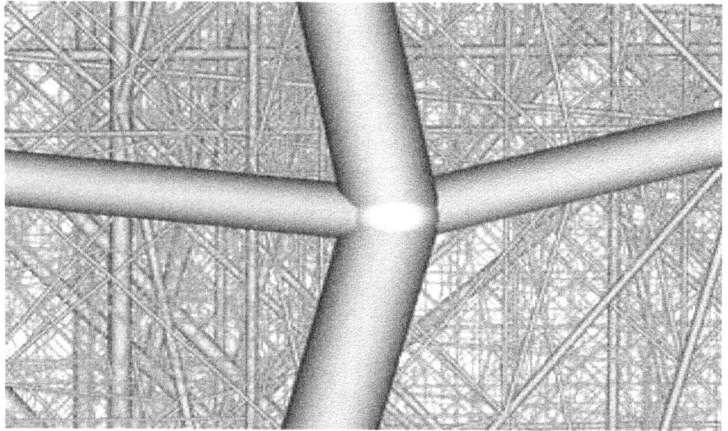

FIG. 6.13 A clear view of segments behind the front equatorial corner.

As the subdivisions increase the number of segments likewise increase. For example, the grid representing the 7th subdivision has 781,250 segments, and the grid representing the 8th subdivision has 3,906,250 (Figure 6.13).

§31 MOVING THE 4-WEB INTO 3-SPACE

The "4" in "4-web" serves to remind us that the 4-web naturally lives inside 4-dimensional space. The fact that the 4-web could be moved into our 3-dimensional vision, however, was not discovered until 2003.[3]

The mathematics behind moving the 4-web into our 3-dimensional visual space is somewhat specialized and involved. Nevertheless, the idea is rather simple — stepping down one dimension, we can have Freddy the penguin with 2-dimensional vision show us the basic idea.

We begin with Figure 6.14 where we see Freddy on the left looking at a *dyad — two line segments joined at one common endpoint*. The point p is in an *extra* dimension — outside of Freddy's viewing plane. He sees neither the point p nor the line segment connecting p to his visual plane.

---

[3]The relevant publication is detailed in §27.

Freddy desires, nevertheless, to move both the point p and the line segment connected to p into his viewing plane.

FIG. 6.14 Freddy ponders moving the point p to the point q.

He knows that p and the line segment connected to p logically exist, so he proceeds as follows: Selecting a point q in his viewing plane (right-side Figure 6.14), Freddy constructs a *triad — three line segments joined at one common endpoint*. He then poses the question:

Can I move both the point p and the segment containing p into my viewing plane so that p occupies the position of q, and the segment containing p occupies the position of the segment containing q ?

While Freddy cannot picture a solution, he knows that two points determine a line and he writes out a formula for the line segment $[p,q]$ that contains both p and q. He then imagines sliding the point p, like a bead on a string, along the line segment toward q. The idea works (Figure 6.15):

FIG. 6.15 Freddy moves (slides bead) p on the line segment $[p,q]$ toward q.

Freddy concludes that *triads in 3-dimensional space can be moved into 2-dimensional space.* He also realizes that the movement was successful because *there were no line segments obstructing the path from q to the center of the triad.*

Analogously, replace Freddy's dyad in 2-dimensional space with a 3-web in 3-dimensional space. Then realize that it is *the holes in the 3-web* that allow the point p and the segments connected to p — all of which exist within the extra (fourth) dimension — to be moved into our 3-dimensional visual space. The point q in our 3-dimensional visual space is selected so that once inside of 3-space p and the segments connected to p do not intersect any of the segments within the 3-web. *It is the existence of the holes in the 3-web that make it possible to move the 4-web into 3-space.*

§32 Comments

Note that the 4-web cell has five corners, each of which is connected to the other four. In addition to the presentation given in this chapter, one may think of the 4-web as part of the so-called 4-*simplex* — a 4-dimensional generalization of the 3-dimensional tetrahedron. The *standard picture* of the five "corners" and ten "edges" of the 4-simplex places the fifth corner *inside* a tetrahedron, making the "fifth corner" a *non-corner*. For those interested in the picture see page 150 in Stillwell's book *Yearning for the Impossible* whose publication details are provided in the first paragraph of §7 above.

In §31 Freddy the penguin, with 2-dimensional vision, motivated the idea of moving the fifth corner of the 4-web cell from the fourth dimension into our 3-dimensional visual space. The idea is also illustrated within fifteen color plates in the author's book *Fractals and Universal Spaces in Dimension Theory*.[4]

Each plate pictures an image within our visual 3-space. The plates begin with a picture of only the basic 3-web cell, which is colored red. Then like buds on a tree trunk, new limbs (line segments) are pictured in a sequence that may be described as "growing of the limbs" until the entire 4-web cell appears in the fifteenth plate.

Keep in mind that a key feature of 4-web grids is that *as the number of subdivisions of a 4-web cell increase without limit, the numbers of line segments in the corresponding 4-web grids also increase without limit.* In the following chapter, when we use segments of 4-web grids to capture 3-sphere dots, we can be sure that we have a sufficient supply of such segments.

Finally, it was Chris Dupilka who created the first graphics that represent various subdivisions of the 4-web. In particular, the graphics appearing in figures 6.3 through 6.13 are either directly or indirectly due to Dupilka, who wrote the necessary algorithms and then encoded the algorithms in the POVRAY software language. In addition, Dupilka created the first video showing a 4-web moving into human view. The video is contained in the supplemental Blu-ray Disc.

---

[4]For publication details of the book see footnote 2 in Chapter 6, or the Bibliography.

CHAPTER 7

The Partial Picture

The idea of a 4-web grid from the fourth dimension was developed in the previous chapter. Such a grid may be moved into human view with its *structure preserved*. While living in the fourth dimension, however, a 4-web grid may be used to capture those points where the grid meets a hyper-sphere (3-sphere). Then these *captured* hyper-sphere points may be moved into human view using the technique for moving the grid into human view. The result? A partial picture of a 3-sphere.

§33 CAMERA POSITION

We begin with an illustration of the camera position relative to the basic 4-web cell (already moved into our visual 3-space).

The gray funnel-sphere object at the top of the illustration is the camera. The camera's lens is located at the small end of the funnel, and the 4-web cell at the bottom is the target. The center of the 4-cell is pictured as the small light-gray sphere, and the black line illustrates the camera line-of-sight. Focusing on the 4-cell target, we see six dark-gray segments that form a 3-web cell (tetrahedron). This 3-cell is fixed within our visual 3-space throughout *the move* of the 4-cell into our visual 3-space. The other four multi-shaded segments of the 4-cell started in the fourth dimension, but are moved into our visual 3-space. Prior to the move, however, we go into the fourth dimension and gray-scale code these four segments:

Each of the multi-shaded segments from the fourth dimension are illustrated by using subsegments that alternate between light and darker shades of gray. Finally, the light-gray line segment from the camera to the 4-cell extends through an equatorial vertex onto the bottom vertex of the 4-cell.

Near-camera views appear in Figure 7.1. The black line depicts camera line-of-sight, and the gray octagonally-shaped area the position of the camera lense. On the right side of Figure 7.1 we see a light-gray line segment connecting the lense to an equatorial corner. The extension of this light-gray line through the equatorial corner is depicted as a darker-gray segment whose terminal endpoint is the south pole. Since the south-pole corner is blocked from camera view by an equatorial corner, we see (left-side graphic) only four of the five corners of the 4-cell.

FIG. 7.1 Near-camera views of a 4-web grid, one showing the south pole.

§34 First picture

Illustrated below we see a gray-scale version of the first partial picture of a 3-sphere produced with the seventh subdivision of a 4-cell. The left-side graphic includes the basic cell (no subdivisions).

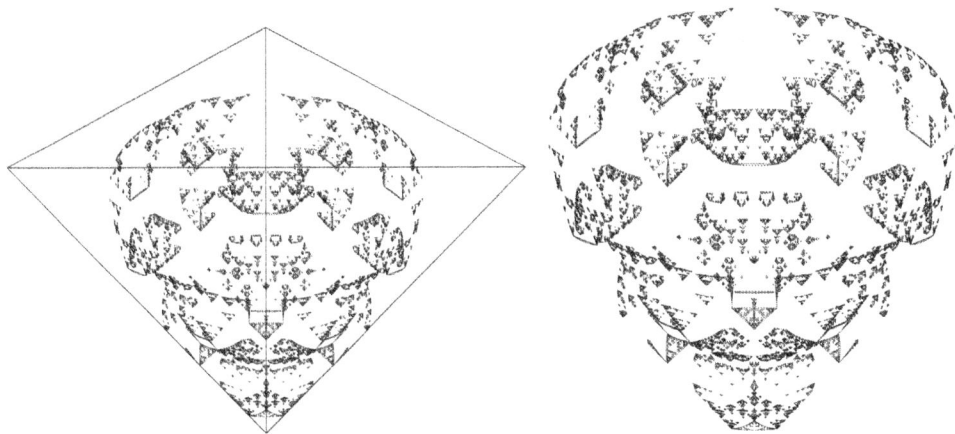

FIG. 7.2 Partial image of a 3-sphere, one with and one without the basic cell.

A larger version of the right-side graphic appears in §27. The left-side graphic requires a few comments. Note again, as discussed above, that only four of the five corners of the 4-web cell appear. Also note that we see only six of the ten line segments connecting the corners of the basic cell. A careful study of the right-side illustration in Figure 7.1 will uncover the reason for the four "missing" segments. Simply put, all four line segments that have the south pole as an endpoint are blocked from camera view.

While this camera and grid configuration may seem ill conceived, the opposite is true — the picture with the grid contains a suppressed number of corners and grid segments, making the cell more transparent which in turn shows more of the partial picture.

§35 WALKING AROUND OUR 3-SPHERE IMAGE

The captured-dots images of the 2-sphere in Figure 5.12 show that a walk around the assemblage of dots is interesting, but a walk around the 3-sphere image is more intriguing. Each view is new, recreating the experience of trying to match pieces of a puzzle whose picture has never been seen.

One example can convey such an experience. If we view the illustration on the right side of Figure 7.2 as something between a *face* and a *skull* then we can label some of the local features. Such labels will aid our navigation and discussion of the partial picture. In particular, the *symmetry* of the human face is the key — *the right half is the mirror image of the left half.*

Viewing the right-side graphic in Figure 7.2 as a *face*, we shall call the bottom area the *chin*. Then the bounding lines that converge toward the chin could be called the *jaw lines*, and the areas adjacent to these lines could be called the *jaws*. Note the symmetry relative to a vertical line that bisects the face. This symmetry includes a symmetry between the two jaws and in the shape of the chin.

Moving up from the chin, we see two nearly-horizontal but pleasantly-pleasing parallel curves that could be viewed as a slight smile. The top pleasantly-pleasing curve will be called the *mouth*.

Just above the mouth we see what we shall call a *mustache*, which has a left and right component — the left half tails off toward the left corner of the mouth and its symmetric right half tails off toward the right corner. The mustache appears as two curved wedge-shaped assemblages of dots, each a symmetric image of the other — among the viewers of the *God's Image?* graphic, some viewed the "mustache" as *spread angel wings*.[1]

Continuing to move up the face, we now see aspects of the graphic generated by points in the fourth dimension. In particular, we see a triangle attached to a rectangle in such a manner that the bottom corner of the triangle points down toward the center of the mustache. The triangle is centered between and just above the two symmetrically-curved wedges that comprise the mustache. We shall call this feature, the rectangle together with the attached triangle, the *nose*.

We know several aspects of human chins, human jaws, human mouths, human mustaches, human noses, and human 2-sphere shaped heads. In particular, from the front-face view we know that the nose is typically centered above the mouth and mustache. This front-view feature agrees with the corresponding front-view feature of our partial picture.

But we also know that as a human head turns, the nose turns with the mouth and the mustache, continuing to mark the center of both. But as is illustrated in Figure 7.3, this is where our visual experience of human facial features deviate from the corresponding features on our assemblage of 3-sphere dots.

---

[1] When the mustache is viewed as "*spread* angel wings" in Figure 7.2 one can also "see" (with some effort) two pairs of "*folded* angel wings": One pair on each side of the vertical central axis — in the right side of Figure 7.2 look at the boundary areas at the level where the "skull" suddenly becomes much wider. The "right-side pair" of "folded wings" appears to be at 45° relative to the vertical, and likewise for the "left-side pair". In fact, all three of these pairs of "angel wings" appear as mustaches when viewed "straight on". The movies within the supplemental Blu-ray Disc clearly expose the "3-fold symmetry" of these "angel wings", which is induced by the 3-fold symmetry of the 4-web cell.

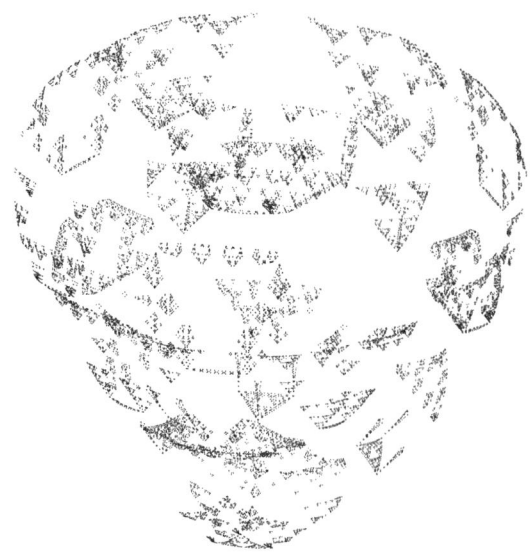

FIG. 7.3 Another view of our assemblage of 3-sphere dots.

Figure 7.3 tells us that our image of the "3-sphere nose" is more centrally located (more near the north-south polar line segment) than is the "3-sphere mustache". Keep in mind, however, that the hyper-sphere graphic may have many interpretations, each depending on the predisposition of the viewer.

Within the context of a human head, as a head turns we see a continuum of images — the back-view hardly looks like the front-view, and the "back-view of a human head" is not viewed as a face of man.

From an artistic view, a continuum of images may be generated by simply moving the camera. One view in particular that this author found interesting is illustrated within the color plates: The new art was generated by placing the camera beneath the "chin" and selecting the midpoint of the top of the "head" as a target point. The illustration clearly demonstrates the 3-fold symmetry of our partial 3-sphere image.

§36 COMMENTS

The supplemental Blu-ray Disc contains five movies of the *God's Image?* graphic, namely,

> *From beyond human vision "4-web meets hyper-sphere" Level-4,*
> ...
> *From beyond human vision "4-web meets hyper-sphere" Level-8*

These movies illustrate how the various levels of a 4-web change the graphic.

Given the central role that the hyper-sphere played in both Dante's circa 1300 AD *Divine Comedy* and the new hyper-sphere art from the fourth dimension, it is ironic that both are based on the number "3".

Indeed, close inspection of the new art shows that *everywhere appearances* of the number "3" via corners of 2-webs, and 2-web triangular shapes — the 4-web is but an extension of the 2-web to the fourth dimension, and thus contains an infinity of triangular shapes. That is, within the new hyper-sphere art the number "3" is basic.

Turning to Dante's *Divine Comedy*, we also find a plethora of "3"s: Consider the following 1970s quote that is parsed from a review of Dante's *Divine Comedy*[2]

> ... The poem, though unique, ... is inspired by the poetry of the Bible and by the Christian wisdom of the Holy Scriptures. Divided into three books, or *cantiche* (treating of Hell, Purgatory, and Paradise—the first composed by 1312, the second by 1315, the third between 1316 and 1321), it is written in Italian vernacular and is composed of 14,233 hendecasyllables in terza rima (a rhyme scheme aba, bcb, cdc, and so on), arranged in 100 cantos (one being a prologue to the entire work, and each a *cantica* having 33 cantos). Thus, the number 3, a symbol of the Trinity, is always present in every part of the work, with its multiples and in its unity. ...

On the following page, we include the *God's Image?* picture using the 5th, 6th, 7th, and 8th subdivisions of the 4-web cell. The size of the spheres used to represent points at various subdivisions is adjusted to improve the quality of the images. For example the size at level-5 is larger than the size at level-8. There is also the question of whether some of the "dots" are larger than the segments of the 4-web grid that produced the dots.

---

[2]The review quoted in this section is part of the review in §13. Section 13 also contains the details of the original reference (Encyclopædia Britannica).

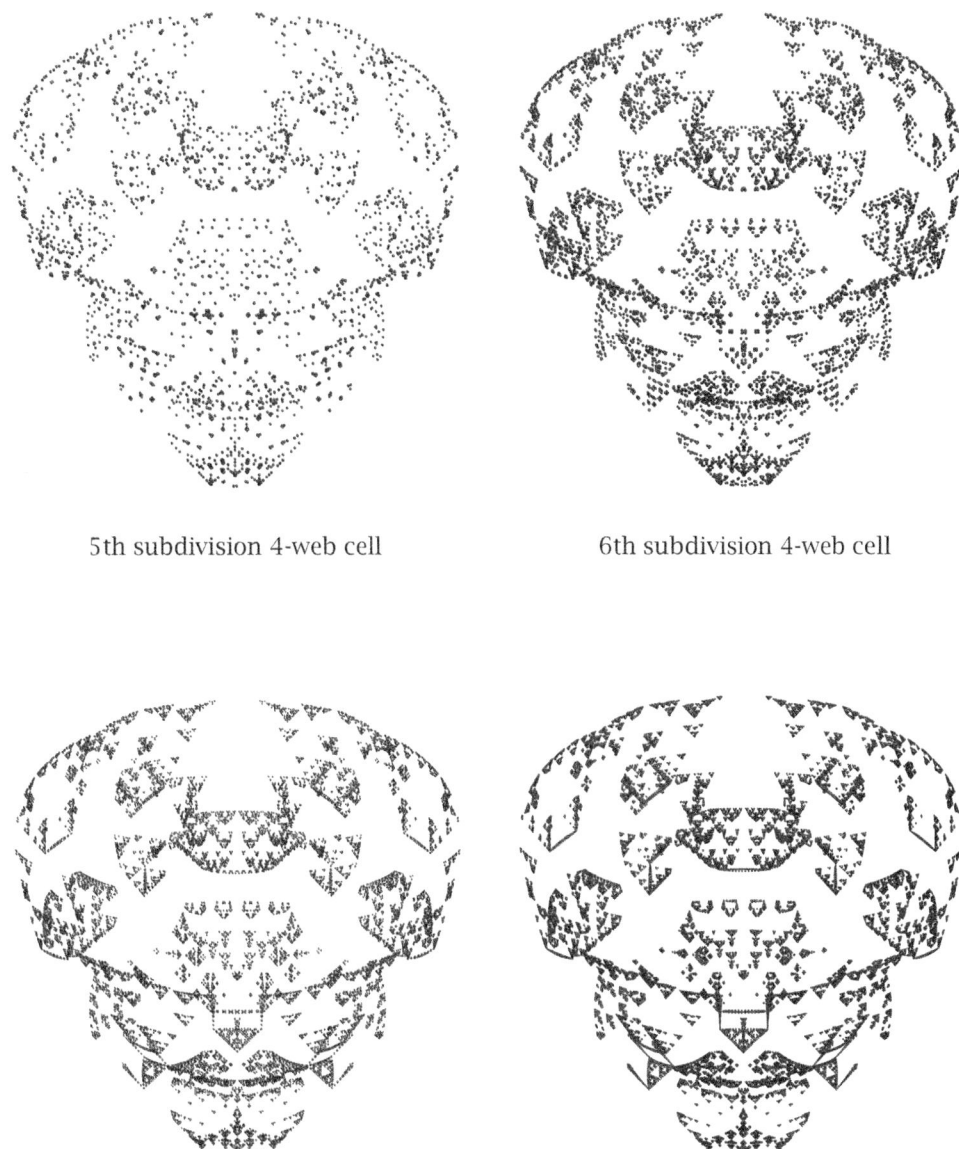

5th subdivision 4-web cell

6th subdivision 4-web cell

7th subdivision 4-web cell

8th subdivision 4-web cell

The 5th subdivision of the 4-web cell has $10 \times 5^5 = 31,250$ line segments, which capture a total of 2,712 points on the hyper-sphere in 4-space. The 6th subdivision has $10 \times 5^6 = 156,250$ segments, which capture a total of 7,104 points on the hyper-sphere. The 7th has $10 \times 5^7 = 781,250$ segments, which capture 17,688 points, and the 8th has $10 \times 5^8 = 3,906,250$ segments, which capture a total of 45,528 points on the hyper-space.

CHAPTER 8

Generating the Hyper-Sphere Art

For those readers who may be interested in generating their own versions of the "God's Image?" art, the computer code that generated the art is presented in this chapter. In addition, the algorithms that underlie the code are also presented. Knowledge of vector and matrix algebra may be helpful.

§37 OVERVIEW OF CHAPTER

The computer code underlying the "God's Image?" art requires input in the form of a camera position, a 4-web grid, and a 3-sphere. By slightly varying these selections one may experience the "God's Image?" art from various perspectives. Such a study is yet to be undertaken, and the task requires an understanding of the relevant computer code. This chapter documents the key details: In §38 4-*Web grids* we find the recursive method of generating 4-web grids in hyper-space; in §39 *Consistency and spheres* we learn that within hyper-space a line segment of a 4-web grid can intersect a 3-sphere in at most two points; in §40 *Points of intersection* we find the algorithm that calculates the points of intersection; and in §41 *From hyper-space into human vision* we provide the algorithm for moving the captured points from hyper-space into human visual 3-space.

§38 4-WEB GRIDS

The recursive algorithm that generates 4-web grids is easily demonstrated by selecting a *level*-0 grid and then using it to create a *level*-1 grid. Human views of a level-0 grid and a level-1 grid appear, respectively, on the left and right sides of Figure 6.3. Note that the five corners of the level-0 grid also appear in the level-1 grid. The level-1 grid is geometrically created from the level-0 grid. The process rests on two steps:

- shrink the level-0 grid by 1/2 toward one of its vertices
- position five copies of the grid produced in step one
  so that each just touches the other four.

Within human view step one and step two appear, respectively, in Figure 6.4 and in Figures 6.5 through 6.8. In general these two steps recursively yield

4-web grids — for any non-negative integer k one may apply the two steps to a level-k grid and thereby create a level-$(k + 1)$ grid.

For an example of the algorithm, we begin with $k = 0$. The key idea for $k = 0$ is *general position*, introduced below with lower-dimensional analogies:

On a line	*Two points* are in general position if they are not equal.
In a plane	*Three points* are in general position if no two are equal, and the three points do not lie on a line.
In human view	*Four points* are in general position if no two are equal, no three lie on a line, and the four points do not lie in a plane.
In hyper-space	*Five points* are in general position if no two are equal, no three lie on a line, no four lie in a plane, and the five points do not lie in a hyper-plane (a human visual space).

Within hyper-space we select five points in general position:

$$\overbrace{(0,0,0,0)}^{\mathbf{u}_0}, \overbrace{(1,0,0,0)}^{\mathbf{u}_1}, \overbrace{(0,1,0,0)}^{\mathbf{u}_2}, \overbrace{(0,0,1,0)}^{\mathbf{u}_3}, \text{ and } \overbrace{(0,0,0,1)}^{\mathbf{u}_4}.$$

These points are the vertices of our level-0 4-web grid in hyper-space. The 10 possible pairs of these five points determine 10 *edges*:

$$[\mathbf{u}_0, \mathbf{u}_1], [\mathbf{u}_0, \mathbf{u}_2], [\mathbf{u}_0, \mathbf{u}_3], [\mathbf{u}_0, \mathbf{u}_4], [\mathbf{u}_1, \mathbf{u}_2],$$
$$[\mathbf{u}_1, \mathbf{u}_3], [\mathbf{u}_1, \mathbf{u}_4], [\mathbf{u}_2, \mathbf{u}_3], [\mathbf{u}_2, \mathbf{u}_4], [\mathbf{u}_3, \mathbf{u}_4].$$

As the notation indicates, each edge is a line segment whose endpoints form a pair of vertices. For the "shrink by 1/2" step, we select vertex $\mathbf{u}_0$ and shrink our level-0 grid by 1/2 toward $\mathbf{u}_0$. The calculations yield level-1 *segments*:

$$\overbrace{[(0,0,0,0),(1/2,0,0,0)]}^{[\frac{1}{2}\mathbf{u}_0, \frac{1}{2}\mathbf{u}_1]}, \overbrace{[(0,0,0,0),(0,1/2,0,0)]}^{[\frac{1}{2}\mathbf{u}_0, \frac{1}{2}\mathbf{u}_2]}, \overbrace{[(0,0,0,0),(0,0,1/2,0)]}^{[\frac{1}{2}\mathbf{u}_0, \frac{1}{2}\mathbf{u}_3]},$$

$$\overbrace{[(0,0,0,0),(0,0,0,1/2)]}^{[\frac{1}{2}\mathbf{u}_0, \frac{1}{2}\mathbf{u}_4]}, \overbrace{[(1/2,0,0,0),(0,1/2,0,0)]}^{[\frac{1}{2}\mathbf{u}_1, \frac{1}{2}\mathbf{u}_2]}, \overbrace{[(1/2,0,0,0),(0,0,1/2,0)]}^{[\frac{1}{2}\mathbf{u}_1, \frac{1}{2}\mathbf{u}_3]},$$

$$\overbrace{[(1/2,0,0,0),(0,0,0,1/2)]}^{[\frac{1}{2}\mathbf{u}_1, \frac{1}{2}\mathbf{u}_4]}, \overbrace{[(0,1/2,0,0),(0,0,1/2,0)]}^{[\frac{1}{2}\mathbf{u}_2, \frac{1}{2}\mathbf{u}_3]}, \overbrace{[(0,1/2,0,0),(0,0,0,1/2)]}^{[\frac{1}{2}\mathbf{u}_2, \frac{1}{2}\mathbf{u}_4]},$$

$$\overbrace{[(0,0,1/2,0),(0,0,0,1/2)]}^{[\frac{1}{2}\mathbf{u}_3, \frac{1}{2}\mathbf{u}_4]}$$

The $\mathbf{u}_0$-part of our level-1 4-web grid.

§38 4-WEB GRIDS

With the first step in our algorithm demonstrated, we note that these ten edges represent 1/5 of the edges that define our level-1 grid. We may think of this part of the level-1 grid as the "$\mathbf{u}_0$-*part*" because our level-0 grid was shrunk toward $\mathbf{u}_0$. The second step provides the remaining four "parts", i.e., the $\mathbf{u}_1$-part, the $\mathbf{u}_2$-part, the $\mathbf{u}_3$-part, and the $\mathbf{u}_4$-part.

We illustrate how the $\mathbf{u}_1$-part is calculated: The level-1 edge $[\frac{1}{2}\mathbf{u}_0, \frac{1}{2}\mathbf{u}_1]$ in the $\mathbf{u}_0$-part is used to create the edge $[\frac{1}{2}\mathbf{u}_0 + \frac{1}{2}\mathbf{u}_1, \frac{1}{2}\mathbf{u}_1 + \frac{1}{2}\mathbf{u}_1]$ in the $\mathbf{u}_1$-part.

In simple terms we move the former to the latter by adding "$\frac{1}{2}\mathbf{u}_1$" to the segment's endpoints:

$$[\tfrac{1}{2}\mathbf{u}_0, \tfrac{1}{2}\mathbf{u}_1] \longrightarrow [\tfrac{1}{2}\mathbf{u}_0 + \tfrac{1}{2}\mathbf{u}_1, \tfrac{1}{2}\mathbf{u}_1 + \tfrac{1}{2}\mathbf{u}_1].$$

Using "components" (numbers), we can express the move as

$$[(0,0,0,0),(\tfrac{1}{2},0,0,0)] \longrightarrow [(\tfrac{1}{2},0,0,0),(1,0,0,0)].$$

This case demonstrates that the move does not change length, i.e., both

$$[\tfrac{1}{2}\mathbf{u}_0, \tfrac{1}{2}\mathbf{u}_1] \quad \text{and} \quad [\tfrac{1}{2}\mathbf{u}_0 + \tfrac{1}{2}\mathbf{u}_1, \tfrac{1}{2}\mathbf{u}_1 + \tfrac{1}{2}\mathbf{u}_1]$$

have length "$\frac{1}{2}$".

To summarize, we present the entire $\mathbf{u}_1$-part of our level-1 grid:

$$\underbrace{[\tfrac{1}{2}(\mathbf{u}_0+\mathbf{u}_1), \tfrac{1}{2}(\mathbf{u}_1+\mathbf{u}_1)]}_{[(.5,0,0,0),(1,0,0,0)]} \; , \; \underbrace{[\tfrac{1}{2}(\mathbf{u}_0+\mathbf{u}_1), \tfrac{1}{2}(\mathbf{u}_2+\mathbf{u}_1)]}_{[(.5,0,0,0),(.5,.5,0,0)]} \; , \; \underbrace{[\tfrac{1}{2}(\mathbf{u}_0+\mathbf{u}_1), \tfrac{1}{2}(\mathbf{u}_3+\mathbf{u}_1)]}_{[(.5,0,0,0),(.5,0,.5,0)]} \; ,$$

$$\underbrace{[\tfrac{1}{2}(\mathbf{u}_0+\mathbf{u}_1), \tfrac{1}{2}(\mathbf{u}_4+\mathbf{u}_1)]}_{[(.5,0,0,0),(.5,0,0,.5)]} \; , \; \underbrace{[\tfrac{1}{2}(\mathbf{u}_1+\mathbf{u}_1), \tfrac{1}{2}(\mathbf{u}_2+\mathbf{u}_1)]}_{[(1,0,0,0),(.5,.5,0,0)]} \; , \; \underbrace{[\tfrac{1}{2}(\mathbf{u}_1+\mathbf{u}_1), \tfrac{1}{2}(\mathbf{u}_3+\mathbf{u}_1)]}_{[(1,0,0,0),(.5,0,.5,0)]} \; ,$$

$$\underbrace{[\tfrac{1}{2}(\mathbf{u}_1+\mathbf{u}_1), \tfrac{1}{2}(\mathbf{u}_4+\mathbf{u}_1)]}_{[(1,0,0,0),(.5,0,0,.5)]} \; , \; \underbrace{[\tfrac{1}{2}(\mathbf{u}_2+\mathbf{u}_1), \tfrac{1}{2}(\mathbf{u}_3+\mathbf{u}_1)]}_{[(.5,.5,0,0),(.5,0,.5,0)]} \; , \; \underbrace{[\tfrac{1}{2}(\mathbf{u}_2+\mathbf{u}_1), \tfrac{1}{2}(\mathbf{u}_4+\mathbf{u}_1)]}_{[(.5,.5,0,0),(.5,0,0,.5)]} \; ,$$

$$\underbrace{[\tfrac{1}{2}(\mathbf{u}_3+\mathbf{u}_1), \tfrac{1}{2}(\mathbf{u}_4+\mathbf{u}_1)]}_{[(.5,0,.5,0),(.5,0,0,.5)]} \; .$$

The $\mathbf{u}_1$-part of our level-1 4-web grid.

The template above for the $\mathbf{u}_1$-part of our level-1 grid serves as a template for the $\mathbf{u}_2$-part — simply replace each instance of "$+\mathbf{u}_1$" with "$+\mathbf{u}_2$" and then do the numerical calculations.

The general pattern should now be clear — the $\mathbf{u}_3$- and $\mathbf{u}_4$- parts may be calculated by replacing, respectively, each instance of "$+\mathbf{u}_1$" in the $\mathbf{u}_1$-part above with "$+\mathbf{u}_3$" and "$+\mathbf{u}_4$". Since each of the five parts yield 10 edges, we see that our level-1 grid contains $10 \times 5 = 50$ line segments.

§39 CONSISTENCY OF SPHERES

Images of a sphere consistently change as the sphere successively moves through lower dimensions.

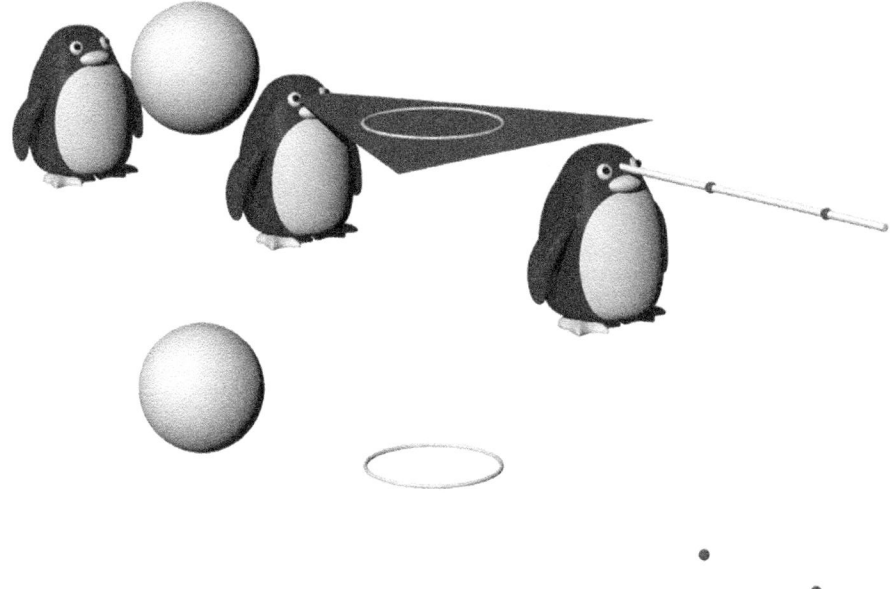

FIG. 8.1 By reducing the dimension of his vision (top row), Freddy creates *images* (bottom row) of a 2-sphere as it moves through successively-lower dimensions.

To illustrate, let us use a 2-sphere moving through Freddy's successive 3-dimensional, 2-dimensional, and 1-dimensional vision (Figure 8.1).

Note the *successively lower-dimensions* pattern

$$S^2, S^1, S^0.$$

In order to add some notation, let us first look at the top of Figure 8.1. Freddy *successively lowers the dimension of his vision* — from a 3-dimensional visual space denoted "π_3", to a 2-dimensional visual space denoted "π_2", to a 1-dimensional visual space denoted "π_1" — and he successively sees the spheres S^2, S^1, and S^0.

But we can remove Freddy from the picture. Start with a 2-*sphere* S^2. Then S^2, *by successively moving through the lower-dimensional visual spaces*,

$$\pi_3, \pi_2, \pi_1, \text{ produces images } S^2, S^1, S^0.$$

This feature of a 2-sphere is consistent with any sphere moving through successively-lower dimensions. For instance, suppose we start with a 6-sphere S^6. Then S^6, by successively moving through lower-dimensional spaces,

$$\pi_7, \pi_6, \pi_5, \pi_4, \pi_3, \pi_2, \pi_1, \text{ produces images } S^6, S^5, S^4, S^3, S^2, S^1, S^0.$$

In general this pattern repeats itself for any sphere: Select a natural number, say n, from the infinite list

$$1, 2, 3, 4, 5, 6, 7, 8, 9, 10, 11, 12, 13, \cdots.$$

Then select the corresponding n-sphere S^n from the corresponding list

$$S^1, S^2, S^3, S^4, S^5, S^6, S^7, S^8, S^9, S^{10}, S^{11}, S^{12}, S^{13}, \cdots.$$

Then S^n, by successively moving through lower-dimensional spaces,

$$\pi_{n+1}, \pi_n, \pi_{n-1}, \pi_{n-2}, \cdots, \pi_4, \pi_3, \pi_2, \pi_1,$$

successively produces the corresponding images

$$S^n, S^{n-1}, S^{n-2}, S^{n-3}, \cdots, S^3, S^2, S^1, S^0.$$

Using these observations and looking at Freddy with one dimensional vision in Figure 8.1 we can surmise that in the hyper-sphere (3-sphere) case, a line segment can meet a 3-sphere in at most two points.

Indeed, suppose a line segment $[\mathbf{p}, \mathbf{q}]$ and a 3-sphere live in hyper-space (4-dimensional space). Then let π_1 denote the line that contains $[\mathbf{p}, \mathbf{q}]$. By selecting a point in hyper-space that is not on the line π_1 we may construct a 2-dimensional plane π_2 in hyper-space. Next, by selecting yet another point in hyper-space that does not lie in the plane π_2, we may construct a hyper-plane (human visual space) π_3 in hyper-space. With these constructs and the arguments given in this section we find that our 3-sphere S^3 partially (perhaps totally) generates a 2-sphere in π_3, a 1-sphere in π_2, and a 0-sphere (two points) in π_1. But since $[\mathbf{p}, \mathbf{q}]$ is inside of π_1, we may surmise that $[\mathbf{p}, \mathbf{q}]$ meets our 3-sphere in at most two points.

§40 POINTS OF INTERSECTION

In this section we provide an algorithm that allows us, with the aid of a computer, to calculate those points in hyper-space that are common to a given 3-sphere and a given 4-web grid.

The algorithm requires some notation. We shall use "$[\mathbf{p}, \mathbf{q}]$" to denote the line segment with endpoints $\mathbf{p} = (p_1, p_2, p_3, p_4)$ and $\mathbf{q} = (q_1, q_2, q_3, q_4)$. And given the *center-point* $\mathbf{c} = (c_1, c_2, c_3, c_4)$ and a *radius* r (a positive number) we have the equation for the 3-sphere *centered at* $\mathbf{c}$ *of radius* r:[1]

$$(x_1 - c_1)^2 + (x_2 - c_2)^2 + (x_3 - c_3)^2 + (x_4 - c_4)^2 = r^2.$$

Turning to a representation of the points $\mathbf{a}_t$ in the line segment $[\mathbf{p}, \mathbf{q}]$ we have, for $\mathbf{m} = \mathbf{q} - \mathbf{p}$,

$$\mathbf{a}_t = t\mathbf{p} + (1-t)\mathbf{q} = \mathbf{q} - t(\mathbf{q} - \mathbf{p}) = \mathbf{q} - t\mathbf{m} \quad (0 \le t \le 1).$$

Note that as t moves from zero to one the corresponding point $\mathbf{a}_t$ moves from $\mathbf{q}$ to $\mathbf{p}$. To use this equation $\mathbf{a}_t = \mathbf{q} - t\mathbf{m}$ of a line segment in the context of the equation of a three sphere we express $\mathbf{a}_t$ and $\mathbf{m}$ in terms of their coordinates:

$$\mathbf{a}_t = (a_1, a_2, a_3, a_4) = (q_1 - tm_1, q_2 - tm_2, q_3 - tm_3, q_4 - tm_4), \text{ where}$$
$$\mathbf{m} = (m_1, m_2, m_3, m_4) = (q_1 - p_1, q_2 - p_2, q_3 - p_3, q_4 - p_4).$$

So in terms of these coordinate representations, the point $\mathbf{a}_t$ in $[\mathbf{p}, \mathbf{q}]$ is also a point in our 3-sphere whenever

$$(a_1 - c_1)^2 + (a_2 - c_2)^2 + (a_3 - c_3)^2 + (a_4 - c_4)^2 = r^2$$

or, upon substitution,

$$(q_1 - tm_1 - c_1)^2 + (q_2 - tm_2 - c_2)^2 + (q_3 - tm_3 - c_3)^2 + (q_4 - tm_4 - c_4)^2 = r^2.$$

A rearrangement of terms then yields

$$(q_1 - c_1 - tm_1)^2 + (q_2 - c_2 - tm_2)^2 + (q_3 - c_3 - tm_3)^2 + (q_4 - c_4 - tm_4)^2 = r^2.$$

We now let $\mathbf{b} = (b_1, b_2, b_3, b_4) = (q_1 - c_1, q_2 - c_2, q_3 - c_3, q_4 - c_4)$, and make a substitution to obtain

$$(b_1 - tm_1)^2 + (b_2 - tm_2)^2 + (b_3 - tm_3)^2 + (b_4 - tm_4)^2 = r^2.$$

Continuing, since each term is squared we also have

$$(tm_1 - b_1)^2 + (tm_2 - b_2)^2 + (tm_3 - b_3)^2 + (tm_4 - b_4)^2 = r^2.$$

Upon squaring each expression and then recognizing the *dot product*

$$\mathbf{m} \cdot \mathbf{b} = m_1 b_1 + m_2 b_2 + m_3 b_3 + m_4 b_4$$

---

[1] The 3-sphere used to generate the "God's Image?" art on the cover is centered at $\mathbf{c} = (.25, .25, .25, .25)$ and has radius $r = .4$.

for example, we arrive at a classical 2nd-degree equation with variable t:

$$t^2 A + tB + C = t^2(\mathbf{m} \cdot \mathbf{m}) - t(2\mathbf{m} \cdot \mathbf{b}) + (\mathbf{b} \cdot \mathbf{b} - r^2) = 0$$

where $A = \mathbf{m} \cdot \mathbf{m}$, $B = -2\mathbf{m} \cdot \mathbf{b}$, and $C = \mathbf{b} \cdot \mathbf{b} - r^2$.

It is well known that such equations have at most two solutions, i.e., there are at most two values of t that provide a solution. So the edge $[\mathbf{p}, \mathbf{q}]$ contains at most two points $\mathbf{a}_t$ that are also points of our 3-sphere.

§41 FROM HYPER-SPACE INTO HUMAN VIEW[2]

With 4-web grids in hyper-space calculated according to the algorithm in §39, and with those points in common to both a 4-web grid and a 3-sphere calculated according to the algorithm in §41, the remaining question is *How do we move these captured points from hyper-space into human view?* In this section we provide the algorithm that answers this question.

The solution involves three equations. Together they move a captured point (x_1, x_2, x_2, x_4) in hyper-space — a point that is contained in both the input 4-web grid and the input 3-sphere — to a corresponding point (y_1, y_2, y_3) within human visual space — the first equation serves to calculate y_1, the second y_2, and the third y_3. The three equations are displayed below:

$$y_1 = 1x_1 + 0x_2 + 0x_3 + (2/3)x_4$$
$$y_2 = 0x_1 + 1x_2 + 0x_3 + (2/3)x_4$$
$$y_3 = 0x_1 + 0x_2 + 1x_3 + (2/3)x_4.$$

For example, suppose the vertex $\mathbf{u}_4 = (0, 0, 0, 1) = (x_1, x_2, x_2, x_4)$ of our 4-web grid in hyper-space is also a point on a 3-sphere. Then this point in hyper-space is moved to the point $(2/3, 2/3, 2/3) = (y_1, y_2, y_3)$ within human view because

$$2/3 = (1 \times 0) + (0 \times 0) + (0 \times 0) + ((2/3) \times 1)$$
$$2/3 = (0 \times 0) + (1 \times 0) + (0 \times 0) + ((2/3) \times 1)$$
$$2/3 = (0 \times 0) + (0 \times 0) + (1 \times 0) + ((2/3) \times 1).$$

§42 COMMENTS

The computer code containing the algorithms presented in this chapter is displayed below. The code was designed for use with the software POV-Ray™ for Windows™.[3]

---

[2] The derivation of the transformation from hyper-space into human visual space may be found in reference [25] and also in the author's book [18].

[3] Version 3.1g.watcom.win 32[Pentium ll optimized], copyright ©1991-1999 by the POV-Team.

When working with the POV-Ray code below one must keep in mind that the POV-Ray coordinate system is left-handed (negatively oriented), while the system for our hyper-space is a positively-oriented system. In particular, the value of the x_3 coordinate of a point (x_1, x_2, x_3, x_4) in our hyper-space has value "$-x_3$" within the POV-Ray system, but the values of "x_1" and "x_2" are the same in both contexts.

The code below was written by Chris Dupilka, one of my students and the first person to discover the God's Image? art. I simply presented my thoughts about using 4-web grids to capture points on a 3-sphere, and one day Chris comes walking into my office with the God's Image? art in hand.

Except for minor adjustments — set aspect ratio to "[1280x1024,AA,0.3]" — the code presented here may be used to approximate the work of art that was later named "God's Image?". The last line of the code includes a file "EdgesI5" that defines a 4-web grid at level-5 as outlined in §38.

The code is presented in typewriter font. The horizontal lines and the non-typewriter font are simply annotations intended to improve the readability of the code.

Camera Position and Lighting in POV-Ray coordinates

```
camera
{
   location < 0 , 0 , -4 >
   sky <1,1,-1>
   look_at < 0.3333 , 0.3333 , -0.3333 >
   angle 20
}
light_source
{
   < 200 , 0 , -200 > color rgb < 0.7 , 0.7 , 0.7 > shadowless
}
light_source
{
   <-200 , 0 , -200 > color rgb < 0.7 , 0.7 , 0.7 > shadowless
}
light_source
{
   < 0 , 200 , -200 > color rgb < 0.7 , 0.7 , 0.7 >
}
```

MACRO Point: Moves coordinates of points in positively-oriented hyper-space into in POV-Ray oriented human visual 3-space. Points in hyper- space are colored according to distance from 3-space.

```
#macro Point (X1, X2, X3, X4)
   /*
   ** This macro plots a point in 3-space.
   */
   #local M11 = 1.0;
   #local M12 = 0.0;
   #local M13 = 0.0;
   #local M14 = 2.0/3.0;

   #local M21 = 0.0;
   #local M22 = 1.0;
   #local M23 = 0.0;
   #local M24 = 2.0/3.0;

   #local M31 = 0.0;
   #local M32 = 0.0;
   #local M33 = -1.0; /* "-" sign for POV-Ray coordinates */
   #local M34 = -2/3; /* "-" sign for POV-Ray coordinates */

   #local Y1 = M11 * X1 + M12 * X2 + M13 * X3 + M14 * X4;
   #local Y2 = M21 * X1 + M22 * X2 + M23 * X3 + M24 * X4;
   #local Y3 = M31 * X1 + M32 * X2 + M33 * X3 + M34 * X4;
   #if (X4 < 0.0)
      #error "Fatal error:  (X4 < 0.0).\n"
   #end
   #if ((0.0 <= X4) & (X4 < 0.2))
      #local RedColor = 1.0;
      #local GreenColor = (X4 - 0.0) * 5.0;
      #local BlueColor = 0.0;
   #end
   #if ((0.2 <= X4) & (X4 < 0.4))
      #local RedColor = 1.0 - (X4 - 0.2) * 5.0;
      #local GreenColor = 1.0;
      #local BlueColor = 0.0;
   #end
   #if ((0.4 <= X4) & (X4 < 0.6))
      #local RedColor = 0.0;
      #local GreenColor = 1.0;
      #local BlueColor = (X4 - 0.4) * 5.0;
   #end
```

```
   #if ((0.6 <= X4) & (X4 < 0.8))
      #local RedColor = 0.0;
      #local GreenColor = 1.0 - (X4 - 0.6) * 5.0;
      #local BlueColor = 1.0;
   #end
   #if ((0.8 <= X4) & (X4 <= 1.0))
      #local RedColor = (X4 - 0.8) * 5.0;
      #local GreenColor = 0.0;
      #local BlueColor = 1.0;
   #end
   #if (1.0 < X4)
      #error "Fatal error:  (1.0 < X4).\n"
   #end

     sphere {< Y1 , Y2, Y3 >, 0.0015 texture {pigment {color rgb <
           RedColor , GreenColor , BlueColor >}}   finish
           {phong 1 phong_size 250 brilliance 1 specular .9}}
#end
```

MACRO Edge: 3-sphere definition in positively-oriented hyper-space. Points of intersection of 3-sphere with segments.

```
#macro Edge (PX1, PX2, PX3, PX4, QX1, QX2, QX3, QX4)
   /*
   ** This macro plots (in 3-space with POV-Ray coordinates)
   ** the points of intersection in hyper-space of a line
   ** segment and a 3-sphere.  The line segment has endpoints
   ** (PX1, PX2, PX3, PX4) and (QX1, QX2, QX3, QX4).
   ** The 3-sphere is defined below.
   */

   #local CenterX1 = 0.25; /* The first coordinate of
         the center of the 3-sphere.  */
   #local CenterX2 = 0.25; /* The second coordinate of
         the center of the 3-sphere.  */
   #local CenterX3 = 0.25; /* The third coordinate of
         the center of the 3-sphere.  */
   #local CenterX4 = 0.25; /* The fourth coordinate of
         the center of the 3-sphere.  */

   #local Radius = 0.4; /* The radius of the 3-sphere.  */
```

```
#local M1 = QX1 - PX1;
#local M2 = QX2 - PX2;
#local M3 = QX3 - PX3;
#local M4 = QX4 - PX4;

#local B1 = PX1 - CenterX1;
#local B2 = PX2 - CenterX2;
#local B3 = PX3 - CenterX3;
#local B4 = PX4 - CenterX4;

#local A = M1 * M1 + M2 * M2 + M3 * M3 + M4 * M4;
#local B = 2.0 * (M1 * B1 + M2 * B2 + M3 * B3 + M4 * B4);
#local C = B1 * B1 + B2 * B2 + B3 * B3 + B4 * B4 -
    Radius * Radius;

#if (A = 0.0)
    #error "Fatal error: (A = 0.0).\n"
#else
    #local Discriminant = B * B - 4.0 * A * C;
    #if (Discriminant < 0) /* There are no solutions to the
        equation (and hence no points of intersection). */
    #end

    #if (Discriminant = 0) /* There is one solution to the
        equation. */

        #local Solution = (0.0 - B) / (2.0 * A);

        #if ((Solution >= 0.0) & (Solution <= 1.0))
            #local SX1 = M1 * Solution + PX1;
            #local SX2 = M2 * Solution + PX2;
            #local SX3 = M3 * Solution + PX3;
            #local SX4 = M4 * Solution + PX4;
            Point (SX1, SX2, SX3, SX4)
        #end
    #end

    #if (Discriminant > 0) /* There are two solutions to the
        equation. */
```

```
            #local Solution1 =
               (0.0 - B + sqrt (Discriminant)) / (2.0 * A);
            #local Solution2 =
               (0.0 - B - sqrt (Discriminant)) / (2.0 * A);

            #if ((Solution1 >= 0.0) & (Solution1 <= 1.0))
               #local SX1 = M1 * Solution1 + PX1;
               #local SX2 = M2 * Solution1 + PX2;
               #local SX3 = M3 * Solution1 + PX3;
               #local SX4 = M4 * Solution1 + PX4;
               Point (SX1, SX2, SX3, SX4)
            #end

            #if ((Solution2 >= 0.0) & (Solution2 <= 1.0))
               #local SX1 = M1 * Solution2 + PX1;
               #local SX2 = M2 * Solution2 + PX2;
               #local SX3 = M3 * Solution2 + PX3;
               #local SX4 = M4 * Solution2 + PX4;
               Point (SX1, SX2, SX3, SX4)
            #end
         #end
      #end
#end

background color rgb < 1 , 1 , 1 >
#include "EdgesI5.pov"
```

Finally, each edge listed in the input file "EdgesI5.pov", which is a list of the edges in a 4-web grid, must be presented in the form "Edge (PX1, PX2, PX3, PX4, QX1, QX2, QX3, QX4)". Each of the eight entries is a number, the first four define one endpoint, the last four the other.

CHAPTER 9

Prelude to Chapters 10 and 11

As detailed in the previous chapter, the "God's Image?" art is generated by computer code that requires input in the form of a camera position, a 4-web grid, and a 3-sphere. Once the art is produced, however, certain aspects of the art may be hidden — *hidden art* may be viewed as shapes or features that go unnoticed, but once explained, become obvious. This prelude is intended to motivate the search (Chapter 10) for, and, the discovery (Chapter 11) of hidden art in the "God's Image?" graphic.

Michelangelo's world-famous fresco *Creation of Adam* provides a classical example of art that contained hidden shapes for nearly five centuries. As a concise presentation of this classic discovery, this chapter should serve to motivate Chapters 10 and 11.

§43 THE FRESCO, CREATION OF ADAM

Michelangelo's world-famous fresco is located on the ceiling of the Sistine Chapel. Many experts place this particular work of art among the most famous images in the world.[1]

It is interesting that nearly five centuries — circa 1511 to 1990 — passed before someone would see a cross-section of the human brain imbedded

---
[1] The file used here to create the illustration of the *Creation of Adam* is a "bmp" version of the freely licensed Wikipedia media file "God2-Sistine_Chapel.png".

within Michelangelo's *Creation of Adam*. That person is Frank Lynn Meshberger. Meshberger saw something totally new, something that once explained became obvious.

But what was Meshberger's history, i.e., what was his predisposition? He was a *medical student*. The details of what Meshberger recognized is presented below as a quote from a BBC web site:[2]

> Viewed purely as a singular work of art, this scene, unnamed by the artist, is arguably one of the greatest masterpieces. Reproduced in countless art books, posters, and postcards, it has become one of the most well-known images in the world. And now, nearly 490 years after Michelangelo painted it, an astonishing theory provides an entirely new perception of the work, and perhaps an insight into the mind of its creator.
>
> *In the fresco traditionally called the 'Creation of Adam', but which might be more aptly titled the 'Endowment of Adam', I believe that Michelangelo encoded a special message.*
> –Frank Meshberger
> *An interpretation of Michelangelo's Creation of Adam Based on Neuroanatomy (JAMA 264:1837-1841, 1990)*
>
> What is usually interpreted from this particular scene is that Adam is not being physically created, but is in the process of receiving something momentous, yet subtle, from the hand of God. Adam's languid posture appears to be one of near mindless repose, whereas the figure of the Creator fairly bristles with energy. The composition enables the viewer almost to perceive the passage of a spark jumping the gap between the outstretched, not quite touching, fingers.
>
> ... If you have an opportunity to view some recent reproductions, or the good fortune to view directly these newly cleaned and restored magnificent frescoes, allow your gaze to linger just a bit longer on *The Creation of Adam*. There are many very good web sites devoted to this particular fresco, and there is a particularly excellent 'virtual tour' of the Sistine frescoes available at the Web Gallery of Art — Sistine Chapel Tour.
>
> It may be that some residual spark persisted over the centuries, to eventually ignite the inspiration of Frank Meshberger, medical student, idly paging through a book about Michelangelo, relaxing after hours of intense study in the neuro-anatomy lab at the Indianapolis School of Medicine. As he gazed at a three-page foldout colour reproduction of the 'Creation of Adam,' he was ...

---

[2] http:/www.bbc.co.uk/dna/h2g2/A681680

> ... *immediately struck by the shape of the image surrounding God and the angels. It was the same thing I had been working with all day! It was the unmistakable outline of the mid-sagittal cross-section of a human brain* – ibid

At this point in the quote, the *image surrounding God* and the *unmistakable outline* of which Frank Meshburger speaks may be illustrated.

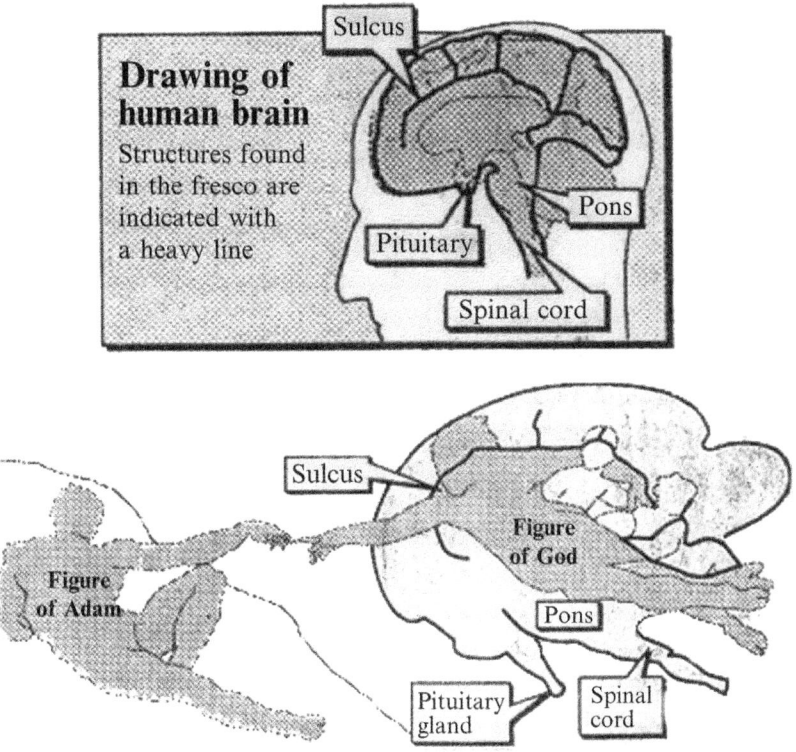

FIG. 9.1 Mid-sagittal cross-section of the human brain (top), and shape of the image surrounding God and the angels (bottom right).[3]

With Figure 9.1 serving as a guide, we return to the BBC web site quote:

> At first, he noticed that the swirling green robe corresponded with the vertebral artery, which follows an irregular path upward toward the pons. Then he noticed the angel's leg extending below the base of the pink outline, that matched the anterior and posterior pituitary. The angel's foot was depicted as bifid, in two parts rather than five toes. Then he noticed the general outline of the sulcus, and the fissure of silvius, which separates the frontal from the parietal lobe.

---

[3] Similar sketches may be found at the BBC web site cited in footnote 2. For these two sketches see "http://www.thecaveonline.com/APEH/michelangelosbrain.html".

Although mildly intrigued at the correspondence between the fresco and the brain cross section, he assumed it was not unknown. Regarding art history trivia as irrelevant to his chosen profession, he went back to his studies, eventually graduating and opening a private practice as an obstetrician/gynecologist. Occasionally, he would casually mention his observation to friends, but it was not until 1988, when the Sistine chapel fresco restoration project was well under way, that it dawned upon him that apparently no one else had ever made a similar observation.

His curiosity piqued anew, Dr Meshberger began researching the life of Michelangelo, and nowhere found any reference to the outline of the human brain on the Sistine Chapel ceiling. He superimposed a transparency of the fresco on a drawing of the sagittal cross section of the brain. The conformity was uncannily, eerily precise:

> *Until I looked through the transparency I didn't realise that one of the angel's backs was the pons, that the legs and hips were the spinal cord ... the knee of the flexed right leg of the angel with the bifid foot represents the transected optic chiasm, the thigh the optic nerve and the leg itself the optic tract ...*
> —ibid

Now let us complete the BBC web-site quote:

> No one could doubt Michelangelo's technical skill, his artistic genius, or his obvious familiarity with gross human anatomy. But how was it possible for even a Renaissance Master to acquire such an intimate knowledge of human neuro-anatomy, particularly the human brain?

So it was Frank's history of medical studies, especially his studies of the human brain, that determined Frank's predisposition to see what he saw. He had to develop his knowledge of the human brain in order to be able to *recognize the pattern*, so to speak. The logic is exemplified in how we think — when you see your wife (or your husband) your brain uses pattern recognition to instantly know who it is that you are looking at. So even though one can prove neither that Michaelangelo knew nor that Michaelangelo did not know of this view of the human brain, Frank Meshberger's "pattern recognition relative to human brains" provided a new, and rather justified, view of the *Creation of Adam*.

§44 Comments

Note that it was Frank Meshberger's predisposed understanding of the "cross-sectional shape of the human brain" that allowed him to *see* the hidden art. Analogously, we may think of Chapter 10 as a presentation that will predispose readers to recognize certain "ellipsoidal shapes" as "images of great 2-spheres". And with such a predisposition in hand, Chapter 11 will highlight those volumes of "dots" in the hyper-sphere graphic that (partially) represent these "ellipsoidal shapes".

CHAPTER 10

Great 2-spheres

The 2-sphere has *great circles*. Examples are illustrated in Figure 10.1 — the "equator" and the two meridians that contain the "north and south poles". Similarly the 3-sphere (hyper-sphere) has *great spheres*.

In this chapter we discuss both a 3-sphere — the one represented as the *God's Image?* graphic — and four of its great 2-spheres. Some knowledge of analytic geometry and matrix algebra is assumed.

§45 Three Great Circles on a 2-sphere

The equation of a unit 2-sphere S^2 centered at the origin of 3-dimensional space induces three pairs of equations. The three pairs correspond to geometrically slicing the sphere with the three planes $x = 0$, $y = 0$, and $z = 0$.

$$x^2 + y^2 + z^2 = 1$$

$$\begin{aligned} y^2 + z^2 &= 1 \\ x &= 0 \end{aligned} \qquad \begin{aligned} x^2 + y^2 &= 1 \\ z &= 0 \end{aligned} \qquad \begin{aligned} x^2 + z^2 &= 1 \\ y &= 0 \end{aligned}$$

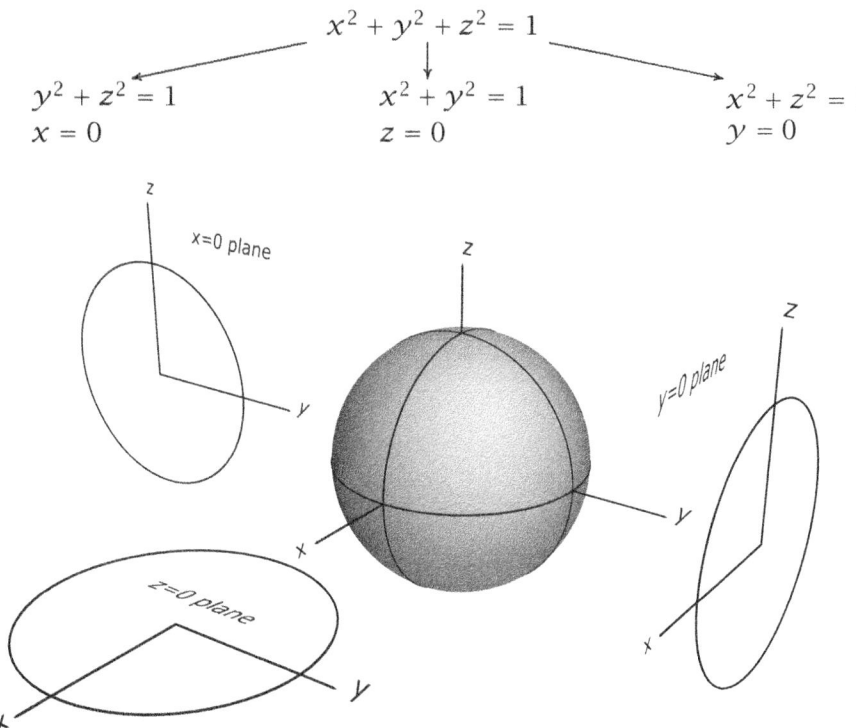

FIG. 10.1 Three great circles which may be called *great 1-spheres*.

Within 3-space, Figure 10.1 illustrates a 2-sphere whose center is the origin $(0,0,0)$ and whose radius is "1" (unity). In the general 2-sphere case, the center can be any point "(a,b,c)" and the radius "r" any positive number. The equations for the general case are listed below

$$(x - a)^2 + (y - b)^2 + (z - c)^2 = r^2$$

$$\begin{array}{c} (y - b)^2 + (z - c)^2 = r^2 \\ x = a \end{array} \qquad \begin{array}{c} (x - a)^2 + (z - c)^2 = r^2 \\ y = b \end{array}$$

$$\begin{array}{c} (x - a)^2 + (y - b)^2 = r^2 \\ z = c \end{array}$$

§46 FOUR GREAT 2-SPHERES

The *unit 3-sphere* S^3 has radius "1" and is centered at the origin of 4-dimensional space. Its points (x, y, z, w) are those whose components satisfy

(1) $$x^2 + y^2 + z^2 + w^2 = 1.$$

A 3-sphere S^3 with radius "r" and center "(a, b, c, d)" consists of those points (x, y, z, w) whose components satisfy

(2) $$(x - a)^2 + (y - b)^2 + (z - c)^2 + (w - d)^2 = r^2.$$

Equation (2) induces four pairs of equations — four great 2-spheres.

$$\begin{array}{c} (x - a)^2 + (z - c)^2 + (w - d)^2 = r^2 \\ y = b \end{array} \qquad \begin{array}{c} (x - a)^2 + (y - b)^2 + (w - d)^2 = r^2 \\ z = c \end{array}$$

$$(x - a)^2 + (y - b)^2 + (z - c)^2 + (w - d)^2 = r^2$$

$$\begin{array}{c} (y - b)^2 + (z - c)^2 + (w - d)^2 = r^2 \\ x = a \end{array} \qquad \begin{array}{c} (x - a)^2 + (y - b)^2 + (z - c)^2 = r^2 \\ w = d \end{array}$$

§47 THE LINEAR TRANSFORMATION L

Light casts shadows on a floor, which is 2-dimensional. Here we step up one dimension, and the transformation "L" casts shadows into 3-dimensional space. The shadow of interest is the *God's Image?* graphic, which is a shadow of those points common to a 4-web grid and the 3-sphere

(3) $$(x - 1/4)^2 + (y - 1/4)^2 + (z - 1/4)^2 + (w - 1/4)^2 = (.4)^2.$$

More precisely, the transformation "L" that creates the shadow in 3-space may be represented by the matrix in

(4) $$\begin{bmatrix} x + (2/3)w \\ y + (2/3)w \\ z + (2/3)w \end{bmatrix} = \begin{bmatrix} 1 & 0 & 0 & 2/3 \\ 0 & 1 & 0 & 2/3 \\ 0 & 0 & 1 & 2/3 \end{bmatrix} \begin{bmatrix} x \\ y \\ z \\ w \end{bmatrix}.$$

Our goal is to use (4) to visualize the "shadows" of the four great 2-spheres obtained by slicing the 3-sphere (3) with hyper-planes $x = 1/4$, $y = 1/4$, $z = 1/4$, and $w = 1/4$.

The shadows take the shapes of ellipsoids. The common center of the four ellipsoids is the image of the center of the 3-sphere, namely

(5) $$\begin{bmatrix} 5/12 \\ 5/12 \\ 5/12 \end{bmatrix} = \begin{bmatrix} 1/4 + (2/3)(1/4) \\ 1/4 + (2/3)(1/4) \\ 1/4 + (2/3)(1/4) \end{bmatrix} = \begin{bmatrix} 1 & 0 & 0 & 2/3 \\ 0 & 1 & 0 & 2/3 \\ 0 & 0 & 1 & 2/3 \end{bmatrix} \begin{bmatrix} 1/4 \\ 1/4 \\ 1/4 \\ 1/4 \end{bmatrix}$$

§48 PLANAR EXAMPLE

In this section we provide an example of how a restriction of L morphs a unit circle into an ellipse. Observe that setting $y = 0$ and $z = 0$ in the (x, y, z, w) vector is equivalent to removing the second and third columns of the matrix in (4). To further simplify our example, we also remove the third row. The result is a (u, v)-planar transformation of the unit circle

(6) $$u^2 + v^2 = 1 \quad \text{and} \quad \begin{bmatrix} x \\ y \end{bmatrix} = \begin{bmatrix} 1 & 2/3 \\ 0 & 2/3 \end{bmatrix} \begin{bmatrix} u \\ v \end{bmatrix}.$$

The expression $f(u, v) = u^2 + v^2$ in (6) is a *quadratic form*. It may be obtained from the *bilinear function*

$$F\left([u_1\ v_1], \begin{bmatrix} u_2 \\ v_2 \end{bmatrix}\right) = [u_1\ v_1] \begin{bmatrix} 1 & 0 \\ 0 & 1 \end{bmatrix} \begin{bmatrix} u_2 \\ v_2 \end{bmatrix}.$$

There are no restrictions on the components u_1, v_1, u_2, and v_2 of F's arguments. When we require, however, that F's arguments be equal, i.e., $u = u_1 = u_2$ and $v = v_1 = v_2$, then

(7) $$f(u, v) = [u\ v] \begin{bmatrix} 1 & 0 \\ 0 & 1 \end{bmatrix} \begin{bmatrix} u \\ v \end{bmatrix}$$

yields the *quadratic form* $f(u,v) = u^2 + v^2$. It turns out that within the context of (6), the shadow of the circle $f(u,v) = 1$ takes the shape of an ellipse $g'(x',y') = 1$.

To see that this is indeed the case, we consider

$$\left. \begin{array}{l} x = u + (2/3)v \\ y = \phantom{u + {}}(2/3)v \end{array} \right\} \iff \left\{ \begin{array}{l} u = x - y \\ v = \phantom{x - {}}(3/2)y \end{array} \right.$$

which, when coupled with $u^2 + v^2 = 1$, provides

(8) $$(x - y)^2 + ((3/2)y)^2 = 1.$$

Expansion of (8) yields the quadratic form

(9) $$g(x,y) = [x\ y] \begin{bmatrix} 1 & -1 \\ -1 & 13/4 \end{bmatrix} \begin{bmatrix} x \\ y \end{bmatrix} = x^2 - 2xy + (13/4)y^2.$$

The form $g(x,y)$ with 2×2 matrix "M" may be reduced to a *canonical* (algebraic) *form* that tells us that the shape of the shadow of the circle $f(x,y) = 1$ is that of an ellipse.

The approach to reducing $x^2 - 2xy + (13/4)y^2 = Ax^2 + Bxy + Cy^2$ to a canonical form rests on removing the mixed term "$-2xy = Bxy$".

The technique involves a *rotation* about the origin of the x-y system through an angle θ. The angle θ may be calculated by solving

$$\tan(2\theta) = \frac{B}{A - C} = \frac{8}{9} \quad \text{so} \quad \theta = 20.816769\ldots \text{ degrees.}$$

Using θ, we define unit vectors

$$\mathbf{e}_1 = \begin{bmatrix} \cos\theta \\ \sin\theta \end{bmatrix} \quad \text{and} \quad \mathbf{e}_2 = \begin{bmatrix} -\sin\theta \\ \cos\theta \end{bmatrix}$$

that specify positive directions of "new" x' and y' axes. These new directions yield the canonical form of (9).

The flow of the reasoning is geometrically illustrated below (reading left-to-right, solid lines/curves indicate the step where the construction occurs):

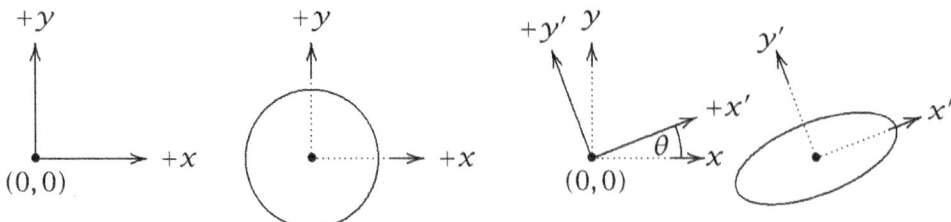

The vectors that define the x' and y' directions are *eigenvectors*. Eigenvectors have associated *eigenvalues* — numbers that define the *major* and *minor* axes, and thus the "shape", of the ellipse. The eigenvalues may be obtained by using the *rotation matrix P* and its *transpose P^T*.

$$P = \begin{bmatrix} \cos\theta & -\sin\theta \\ \sin\theta & \cos\theta \end{bmatrix} \quad \text{and} \quad P^T = \begin{bmatrix} \cos\theta & \sin\theta \\ -\sin\theta & \cos\theta \end{bmatrix}.$$

The columns of P are eigenvectors, and as a transformation matrix, we have

$$\begin{bmatrix} x \\ y \end{bmatrix} = \begin{bmatrix} \cos\theta & -\sin\theta \\ \sin\theta & \cos\theta \end{bmatrix} \begin{bmatrix} x' \\ y' \end{bmatrix} \quad \text{which is denoted} \quad \mathbf{x} = P\mathbf{x}'$$

where the notations "$\mathbf{x}$" and "$\mathbf{x}'$" represent, respectively, vectors whose components reference the (x, y)- and (x', y')-coordinate systems. As for P^T we have $\mathbf{x}' = P^T \mathbf{x}$.

These equations involving P and P^T are keys to calculating the form $\lambda_1 x'^2 + \lambda_2 y'^2$ with no "mixed term":

$$[x \ y] \begin{bmatrix} 1 & -1 \\ -1 & 13/4 \end{bmatrix} \begin{bmatrix} x \\ y \end{bmatrix} = \left(P \begin{bmatrix} x' \\ y' \end{bmatrix}\right)^T \begin{bmatrix} 1 & -1 \\ -1 & 13/4 \end{bmatrix} P \begin{bmatrix} x' \\ y' \end{bmatrix}$$

$$= [x' \ y'] P^T \begin{bmatrix} 1 & -1 \\ -1 & 13/4 \end{bmatrix} P \begin{bmatrix} x' \\ y' \end{bmatrix}$$

$$= [x' \ y'] \begin{bmatrix} \lambda_1 & 0 \\ 0 & \lambda_2 \end{bmatrix} \begin{bmatrix} x' \\ y' \end{bmatrix}.$$

The diagonal 2×2 matrix D above contains the *eigenvalues* λ_1 and λ_2 of the matrix M of $g(x, y)$ (see (9)). To calculate the eigenvalues, one may use a hand calculator with $\theta = 20.816769$ degrees — first calculate the values of the entries of P and P^T, and then perform the multiplications involving P and P^T displayed above. Approximate values are

$$\lambda_1 = .619800677654\ldots \quad \text{and} \quad \lambda_2 = 3.63019932236\ldots \ .$$

The quadratic form $g'(x', y') = (\mathbf{x}')^T D \mathbf{x}' = \lambda_1 x'^2 + \lambda_2 y'^2$ without a "mixed term" is then reduced to the *canonical form of an ellipse* — let $a = 1/\sqrt{\lambda_1} > 0$ and $b = 1/\sqrt{\lambda_2} > 0$, and then express $\lambda_1 x'^2 + \lambda_2 y'^2 = 1$ as

$$\frac{x'^2}{a^2} + \frac{y'^2}{b^2} = 1.$$

The major axis has length $2a$ and minor axis length $2b$.

§49 Great 2-sphere images

The great 2-spheres of our 3-sphere (3) are obtained by a *scaling* followed by a *translation* of the great 2-spheres of the unit 3-sphere. These "operations" are illustrated in Figure 10.2.

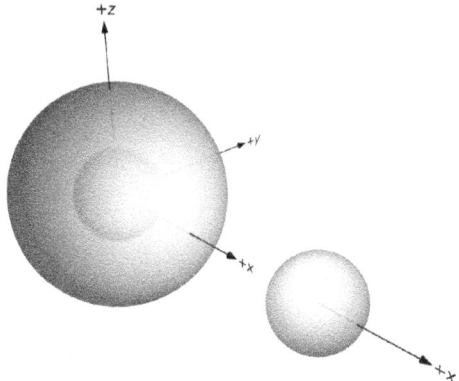

FIG. 10.2 *Scaling* (shrinking in this case) the unit 2-sphere (left side) toward its center yields the "smaller" sphere inside the larger sphere. The smaller sphere is then *translated* along the x axis.

These techniques are applied in the following sections. We begin with four great 2-spheres of the unit 3-sphere.

(10)
$$\begin{cases} y^2 + z^2 + w^2 = 1 \\ x = 0 \end{cases} \quad \begin{cases} x^2 + z^2 + w^2 = 1 \\ y = 0 \end{cases}$$
$$\begin{cases} x^2 + y^2 + w^2 = 1 \\ z = 0 \end{cases} \quad \begin{cases} x^2 + y^2 + z^2 = 1 \\ w = 0 \end{cases}.$$

For each of these spheres, we apply the L transformation (4):

(11)
$$\begin{bmatrix} 1 & 0 & 0 & 2/3 \\ 0 & 1 & 0 & 2/3 \\ 0 & 0 & 1 & 2/3 \end{bmatrix} \begin{bmatrix} 0 \\ y \\ z \\ w \end{bmatrix}, \quad \begin{bmatrix} 1 & 0 & 0 & 2/3 \\ 0 & 1 & 0 & 2/3 \\ 0 & 0 & 1 & 2/3 \end{bmatrix} \begin{bmatrix} x \\ 0 \\ z \\ w \end{bmatrix},$$

$$\begin{bmatrix} 1 & 0 & 0 & 2/3 \\ 0 & 1 & 0 & 2/3 \\ 0 & 0 & 1 & 2/3 \end{bmatrix} \begin{bmatrix} x \\ y \\ 0 \\ w \end{bmatrix}, \quad \begin{bmatrix} 1 & 0 & 0 & 2/3 \\ 0 & 1 & 0 & 2/3 \\ 0 & 0 & 1 & 2/3 \end{bmatrix} \begin{bmatrix} x \\ y \\ z \\ 0 \end{bmatrix}.$$

The impact of $x = 0$ amounts to removing the first column of our transformation matrix, of $y = 0$ removing the second, of $z = 0$ the third, and $w = 0$ the fourth. Thus we may reduce (11) as follows:

$$\begin{bmatrix} 0 & 0 & 2/3 \\ 1 & 0 & 2/3 \\ 0 & 1 & 2/3 \end{bmatrix}_{x=0} \begin{bmatrix} y \\ z \\ w \end{bmatrix}, \quad \begin{bmatrix} 1 & 0 & 2/3 \\ 0 & 0 & 2/3 \\ 0 & 1 & 2/3 \end{bmatrix}_{y=0} \begin{bmatrix} x \\ z \\ w \end{bmatrix},$$

$$\begin{bmatrix} 1 & 0 & 2/3 \\ 0 & 1 & 2/3 \\ 0 & 0 & 2/3 \end{bmatrix}_{z=0} \begin{bmatrix} x \\ y \\ w \end{bmatrix}, \quad \begin{bmatrix} 1 & 0 & 0 \\ 0 & 1 & 0 \\ 0 & 0 & 1 \end{bmatrix}_{w=0} \begin{bmatrix} x \\ y \\ z \end{bmatrix}.$$

Each of these column-reduced matrices is square and non-singular. It follows that the L-transformation (4) is faithful (one-one) on each of our four unit spheres (10). Notice that in the $w = 0$ case the unit 2-sphere $x^2 + y^2 + z^2 = 1$ is *invariant* (unchanged).

What is L doing to points in 4-space? To partially answer this question let us look at the graphic below.

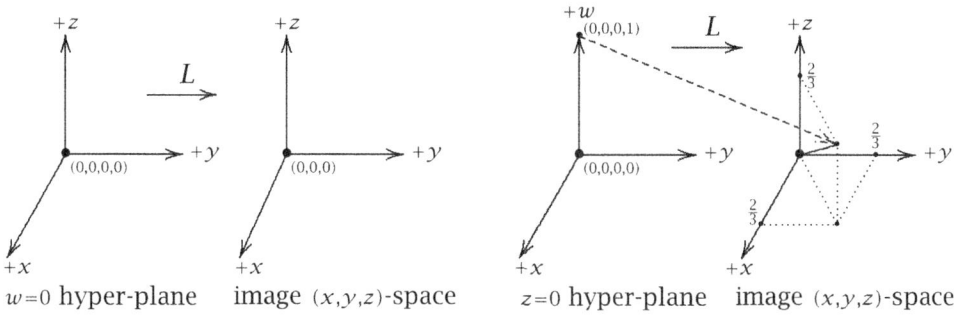

$w=0$ hyper-plane   image (x,y,z)-space $z=0$ hyper-plane   image (x,y,z)-space

On the left side we see L restricted to the $w = 0$ hyper-plane. In this case the $(x,0,0,0)$, $(0,y,0,0)$, and $(0,0,z,0)$ points move, respectively, to $(x,0,0)$, $(0,y,0)$, and $(0,0,z)$. This fact follows because the first three columns of L define the 3×3 identity matrix. In contrast, on the right side of the graphic we see that L moves the $(0,0,0,1)$ point to $(2/3, 2/3, 2/3)$ while the x and y axes are again, in the same sense as the left-side graphic, invariant. It may add insight to think about this "move" physically — view the origins $(0,0,0)$ and $(0,0,0,0)$ as the same point, and then think of someone in 3-space reaching into 4-space, grabbing the w axis, and then, using the origin "$(0,0,0,0) = (0,0,0)$" as a pivot point, rotating the w-axis around the origin and into 3-space where it coalesces with the line through $(0,0,0)$ and $(2/3, 2/3, 2/3)$. During the rotation, however, we need to view the w-axis as an infinitely-straight rubber band that experiences a scaling (stretching) so that whenever it coalesces with the line in 3-space, the point $(0,0,0,1)$ on the rubber band coalesces with $(2/3, 2/3, 2/3)$. This intuitive physical view is consistent with the "action" of L.

§50 THE $w = 1/4$ GREAT 2-SPHERE

In this section we shall use "(x, y, z, w)" as a mnemonic for the 4-space domain of L and "(x', y', z')" as a mnemonic for the image space of L.

The great 2-sphere of interest is specified by the equations

$$\text{(12)} \quad \begin{cases} (x - 1/4)^2 + (y - 1/4)^2 + (z - 1/4)^2 = (.4)^2 \\ w = 1/4. \end{cases}$$

The L-image of the sphere (12) consists of all points (x', y', z') whose components satisfy

$$\text{(13)} \quad \begin{bmatrix} x' \\ y' \\ z' \end{bmatrix} = \begin{bmatrix} 1 & 0 & 0 & 2/3 \\ 0 & 1 & 0 & 2/3 \\ 0 & 0 & 1 & 2/3 \end{bmatrix} \begin{bmatrix} x \\ y \\ z \\ 1/4 \end{bmatrix} \quad \text{for some point } (x, y, z, 1/4) \text{ whose components satisfy (12).}$$

The great 2-sphere of interest and its L image are pictured in Figure 10.3.

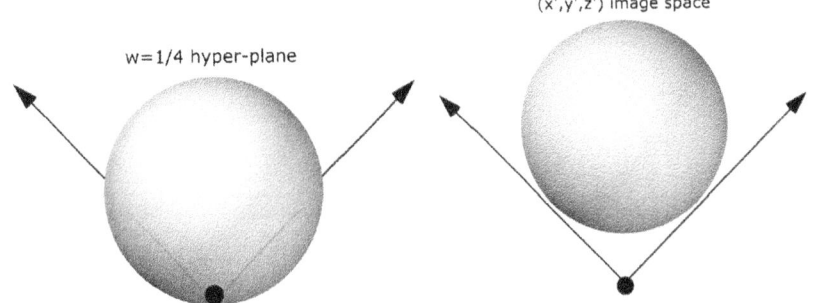

FIG. 10.3 The left-side scene is from inside the $w = 1/4$ hyper-plane. The scene contains the view of the great 2-sphere (12) from a camera located at $(0, 0, 5, 1/4)$ with target $(1/3, 1/3, 1/3, 1/4)$. The right-side scene is the (x', y', z') image space specified by (13). The scene contains the view of the L-image sphere (15) from a camera located at $(0, 0, 5)$ with target $(1/3, 1/3, 1/3)$.

From (13) we have

$$\text{(14)} \quad \left. \begin{array}{l} x' = x + (2/3)(1/4) \\ y' = y + (2/3)(1/4) \\ z' = z + (2/3)(1/4) \end{array} \right\} \iff \begin{cases} x = x' - (2/12) \\ y = y' - (2/12) \\ z = z' - (2/12). \end{cases}$$

And from the right side of (14), upon substituting $x' - (2/12)$, $y' - (2/12)$, and $z' - (2/12)$, respectively, for x, y, and z in (12), we arrive at an equation that specifies the L-image sphere.

$$\text{(15)} \quad (x' - 5/12)^2 + (y' - 5/12)^2 + (z' - 5/12)^2 = (.4)^2.$$

§51 THE $z = 1/4$ GREAT 2-SPHERE

As in §50 we continue to use "(x, y, z, w)" and "(x', y', z')", respectively, as mnemonics for 4-space and 3-space. In addition, however, we shall also use "(x', y', z')" to label points relative to a basis of eigenvectors. Each use of $(x', y,' z')$ should be clear from the context.

The great 2-sphere of interest consists of all points $(x, y, z, w) = (x, y, 1/4, w)$ whose components satisfy

(16) $$\begin{cases} (x - 1/4)^2 + (y - 1/4)^2 + (w - 1/4)^2 = (.4)^2 \\ z = 1/4 \end{cases}$$

And the ellipsoid in Figure 10.4 consists of all *image* points (x', y', z') whose components satisfy

(17) $$\begin{bmatrix} x' \\ y' \\ z' \end{bmatrix} = \begin{bmatrix} 1 & 0 & 0 & 2/3 \\ 0 & 1 & 0 & 2/3 \\ 0 & 0 & 1 & 2/3 \end{bmatrix} \begin{bmatrix} x \\ y \\ 1/4 \\ w \end{bmatrix} \quad \text{for some point } (x, y, 1/4, w) \text{ whose components satisfy (16).}$$

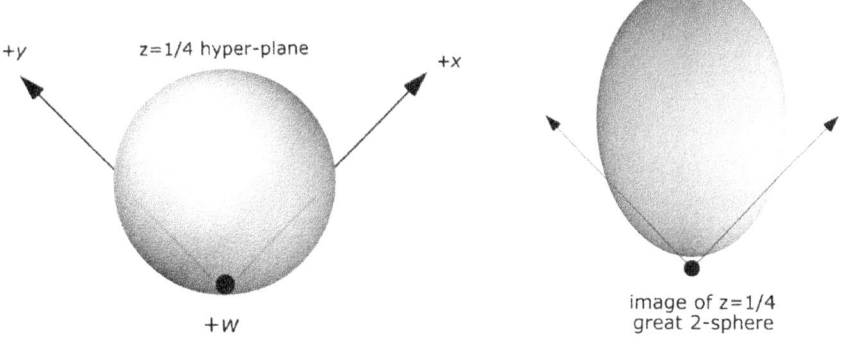

FIG. 10.4 The left-side scene is from inside the $z = 1/4$ hyper-plane. The scene contains the view of the great 2-sphere (16) from a camera located at $(0, 0, 1/4, 5)$ with target $(1/3, 1/3, 1/4, 1/3)$. The right-side scene is the image space specified by (17). The scene contains the view of the L-image ellipsoid (25) from a camera located at $(0, 0, 5)$ with target $(1/3, 1/3, 1/3)$.

The algebraic equations required to explicitly define the ellipsoid is the focus of the remainder of this section. We begin with a unit 2-sphere consisting of all points $(u, v, s, t) = (u, v, 0, t)$ whose components satisfy

(18) $$\begin{cases} u^2 + v^2 + t^2 = 1 \\ s = 0 \end{cases}$$

The L image of this unit 2-sphere (18) is then specified by the points (x, y, z) whose components satisfy

(19) $$\begin{bmatrix} x \\ y \\ z \end{bmatrix} = \begin{bmatrix} 1 & 0 & 0 & 2/3 \\ 0 & 1 & 0 & 2/3 \\ 0 & 0 & 1 & 2/3 \end{bmatrix} \begin{bmatrix} u \\ v \\ 0 \\ t \end{bmatrix}$$ for some point $(u, v, 0, t)$ whose components satisfy (18).

From equation (19) we have

$$\left. \begin{array}{l} x = u + (2/3)t \\ y = v + (2/3)t \\ z = (2/3)t \end{array} \right\} \iff \left\{ \begin{array}{l} u = x - z \\ v = y - z \\ t = (3/2)z \end{array} \right.$$

which, when coupled with $u^2 + v^2 + t^2 = 1$, provides

(20) $$(x - z)^2 + (y - z)^2 + ((3/2)z)^2 = 1.$$

Expansion of (20) yields the quadratic form

(21) $$\begin{aligned} g(x, y, z) &= x^2 + y^2 + (17/4)z^2 - 2xz - 2xy \\ &= [x\ y\ z] \begin{bmatrix} 1 & 0 & -1 \\ 0 & 1 & -1 \\ -1 & -1 & 17/4 \end{bmatrix} \begin{bmatrix} x \\ y \\ z \end{bmatrix}. \end{aligned}$$

The eigenvalues of the *matrix M of* $g(x, y, z)$ are calculated by solving the equation $|\lambda I - M| = 0$ where the vertical bars indicate *determinant*, i.e.,

$$\begin{vmatrix} (\lambda - 1) & 0 & 1 \\ 0 & (\lambda - 1) & 1 \\ 1 & 1 & (\lambda - 17/4) \end{vmatrix} = (\lambda - 1)\bigl[(\lambda - 1)(\lambda - 17/4) - 2\bigr] = 0.$$

The obvious root is $\lambda_1 = 1$, and the other two roots are specified by an application of the quadratic formula. The three solutions (eigenvalues) are

$$\lambda_1 = 1, \qquad \lambda_2 = \frac{21 + 3\sqrt{33}}{8} \approx 4.77921\ldots, \qquad \lambda_3 = \frac{21 - 3\sqrt{33}}{8} \approx .470789\ldots\ .$$

Each of these eigenvalues has a corresponding eigenvector, and these eigenvectors may be calculated using

$$\begin{bmatrix} 1 & 0 & -1 \\ 0 & 1 & -1 \\ -1 & -1 & 17/4 \end{bmatrix} \begin{bmatrix} x \\ y \\ z \end{bmatrix} = \lambda \begin{bmatrix} x \\ y \\ z \end{bmatrix}.$$

§51 THE z = 1/4 GREAT 2-SPHERE

The matrix equation above provides three equations in three unknowns:

$$\begin{aligned} x + 0 - z &= \lambda x \\ 0 + y - z &= \lambda y \\ -x - y + \tfrac{17}{4} z &= \lambda z. \end{aligned}$$

The first and second equations yield $z = (1 - \lambda)x$ and $z = (1 - \lambda)y$.

If $\lambda = \lambda_1 = 1$, then $z = 0$ and the third equation shows that $y = -x$.

If $\lambda \neq \lambda_1 = 1$, then $x = y$ and $z = (1 - \lambda)x$.

So *an eigenvector associated with* λ_1 may be formulated using $\lambda = \lambda_1 = 1$ and $x = 1$. The corresponding *unit eigenvector* $\mathbf{e}_1$ appears on the right side below:

$$\begin{bmatrix} 1 \\ -1 \\ 0 \end{bmatrix} \text{ and } \frac{1}{\sqrt{2}} \begin{bmatrix} 1 \\ -1 \\ 0 \end{bmatrix} = \begin{bmatrix} 1/\sqrt{2} \\ -1/\sqrt{2} \\ (1-\lambda_1)/\sqrt{2} \end{bmatrix} = \begin{bmatrix} 1/\sqrt{2 + (1-\lambda_1)^2} \\ -1/\sqrt{2 + (1-\lambda_1)^2} \\ (1-\lambda_1)/\sqrt{2 + (1-\lambda_1)^2} \end{bmatrix}$$

The far-right *form* of the unit eigenvector $\mathbf{e}_1$ undergoes a sign change in the $\mathbf{e}_2$ and $\mathbf{e}_3$ cases — the sign of the middle component changes:

$$\mathbf{e}_2 = \begin{bmatrix} 1/\sqrt{2 + (1-\lambda_2)^2} \\ 1/\sqrt{2 + (1-\lambda_2)^2} \\ (1-\lambda_2)/\sqrt{2 + (1-\lambda_2)^2} \end{bmatrix} \text{ and } \mathbf{e}_3 = \begin{bmatrix} 1/\sqrt{2 + (1-\lambda_3)^2} \\ 1/\sqrt{2 + (1-\lambda_3)^2} \\ (1-\lambda_3)/\sqrt{2 + (1-\lambda_3)^2} \end{bmatrix}.$$

Using the "dot product", we easily calculate that $\mathbf{e}_1 \cdot \mathbf{e}_2 = 0$ and $\mathbf{e}_1 \cdot \mathbf{e}_3 = 0$. Also $\mathbf{e}_2 \cdot \mathbf{e}_3 = 0$ because $(1-\lambda_2)(1-\lambda_3) = -2$. Thus, these three eigenvectors form an *orthonormal basis* for an (x', y', z') 3-space. We may picture these eigenvectors in the context of the (x, y, z)-image space specified by (19). And parallel to §48 where the image of a unit circle is transformed into an ellipse within an (x', y') system, the (x', y', z') system may be used to picture the ellipsoidal image of the $(u, v, 0, w)$ 2-sphere (18).

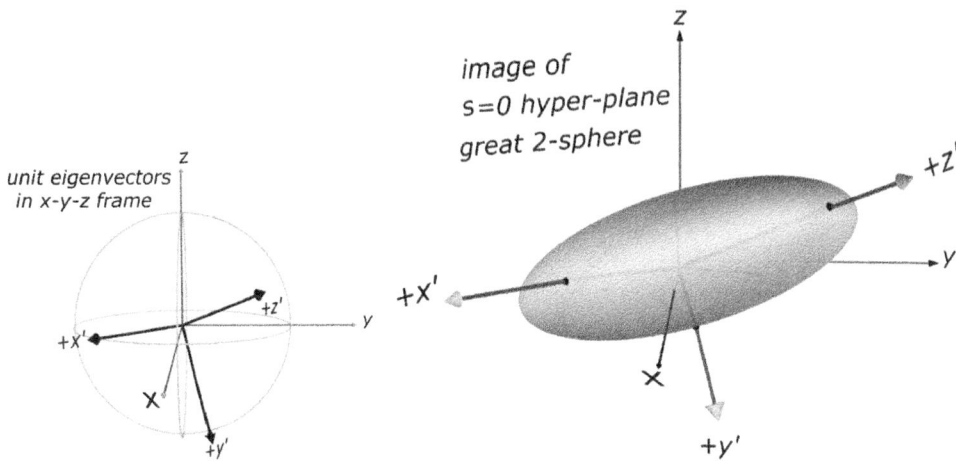

FIG. 10.5 Both scenes show an (x, y, z) system that represents the "L-image" 3-space specified by (19). The left-side scene contains the unit eigenvectors $\mathbf{e}_1$, $\mathbf{e}_2$, and $\mathbf{e}_3$ that define, respectively, the $(+x')$-, $(+y')$-, and $(+z')$-directions that form the (x', y', z') system. The right-side scene contains an ellipsoid centered at $(0, 0, 0)$ whose axes align along the eigenvectors. The ellipsoid is the "L image" (19) of the unit $(u, v, 0, t)$ 2-sphere (18).

To show that the quadratic form $x^2 + y^2 + (17/4)z^2 - 2xz - 2xy$ with "mixed terms" in the (x, y, z) frame equals the quadratic form $\lambda_1 x'^2 + \lambda_2 y'^2 + \lambda_3 z'^2$ with "no mixed terms" in the (x', y', z') frame, we define a matrix P that transforms the (x', y', z') points to the (x, y, z) points:

$$P = \begin{bmatrix} 1/\sqrt{2 + (1 - \lambda_1)^2} & 1/\sqrt{2 + (1 - \lambda_2)^2} & 1/\sqrt{2 + (1 - \lambda_3)^2} \\ -1/\sqrt{2 + (1 - \lambda_1)^2} & 1/\sqrt{2 + (1 - \lambda_2)^2} & 1/\sqrt{2 + (1 - \lambda_3)^2} \\ (1 - \lambda_1)/\sqrt{2 + (1 - \lambda_1)^2} & (1 - \lambda_2)/\sqrt{2 + (1 - \lambda_2)^2} & (1 - \lambda_3)/\sqrt{2 + (1 - \lambda_3)^2} \end{bmatrix}.$$

Note that the first, second, and third columns of P are, respectively, the eigenvectors $\mathbf{e}_1$, $\mathbf{e}_2$, and $\mathbf{e}_3$. Since P has orthonormal columns it is an *orthogonal matrix*. In vector notation we have $\mathbf{x} = P\mathbf{x}'$, and in reverse the *transpose* P^T of P satisfies $\mathbf{x}' = P^T\mathbf{x}$, i.e., P^T is the inverse of P. So, e.g., $\mathbf{x}' = P^T\mathbf{e}_1$ is

$$\begin{bmatrix} 1 \\ 0 \\ 0 \end{bmatrix} = \begin{bmatrix} 1/\sqrt{2} & -1/\sqrt{2} & 0 \\ 1/\sqrt{2+(1-\lambda_2)^2} & 1/\sqrt{2+(1-\lambda_2)^2} & (1-\lambda_2)/\sqrt{2+(1-\lambda_2)^2} \\ 1/\sqrt{2+(1-\lambda_3)^2} & 1/\sqrt{2+(1-\lambda_3)^2} & (1-\lambda_3)/\sqrt{2+(1-\lambda_3)^2} \end{bmatrix} \begin{bmatrix} 1/\sqrt{2} \\ -1/\sqrt{2} \\ 0 \end{bmatrix}$$

where $\mathbf{x}' = [1, 0, 0]^T = 1\mathbf{e}_1 + 0\mathbf{e}_2 + 0\mathbf{e}_3$, and $\mathbf{x} = \mathbf{e}_1 = [1/\sqrt{2}, -1/\sqrt{2}, 0]^T$ has components relative to the (x, y, z) system.

§51   THE $z = 1/4$ GREAT 2-SPHERE

The matrix P along with the matrix M of (21) yield (22) below: Indeed,

$$[x\ y\ z]\begin{bmatrix} 1 & 0 & -1 \\ 0 & 1 & -1 \\ -1 & -1 & 17/4 \end{bmatrix}\begin{bmatrix} x \\ y \\ z \end{bmatrix} = \mathbf{x}^T M \mathbf{x} = (P\mathbf{x}')^T M (P\mathbf{x}')$$

$$= (\mathbf{x}')^T (P^T M P) (\mathbf{x}')$$

$$= [x'\ y'\ z']\begin{bmatrix} \lambda_1 & 0 & 0 \\ 0 & \lambda_2 & 0 \\ 0 & 0 & \lambda_3 \end{bmatrix}\begin{bmatrix} x' \\ y' \\ z' \end{bmatrix}.$$

So the quadratic form $g(x, y, z)$ specified in (21) equates to the (x', y', z') representation on the right side of the following equality:

$$x^2 + y^2 + (17/4)z^2 - 2xz - 2xy = \lambda_1 x'^2 + \lambda_2 y'^2 + \lambda_3 z'^2.$$

From the restriction (20) we also have $g(x, y, z) = 1$. So

(22) $$\frac{x'^2}{(1/\sqrt{\lambda_1})^2} + \frac{y'^2}{(1/\sqrt{\lambda_2})^2} + \frac{z'^2}{(1/\sqrt{\lambda_3})^2} = 1.$$

Equation (22) is that of the ellipsoid pictured in Figure 10.5. Its center is $(0, 0, 0)$ and its axes lie along the x'-, y'-, and z'-axes. The x' axis of the ellipsoid has length $2/\sqrt{\lambda_1}$, its y' axis has length $2/\sqrt{\lambda_2}$, and its z' axis has length $2/\sqrt{\lambda_3}$.

This ellipsoid (22) pictured in Figure 10.5 is key to picturing the L image (ellipsoid in Figure 10.4) of the $z = 1/4$ great 2-sphere (16). That is, we shall scale (shrink) the ellipsoid in Figure 10.5 by a factor of $(.4)$, and then translate this smaller ellipsoid to obtain the ellipsoid of Figure 10.4.

In particular, the method is an application of the properties of a *linear transformation*: If $\mathbf{p}$ and $\mathbf{q}$ are vectors and α is a number, then

$$L(\alpha \mathbf{p} + \mathbf{q}) = \alpha L(\mathbf{p}) + L(\mathbf{q}).$$

To apply this feature of L, we introduce some notation: Let "S^2" denote the 2-sphere $u^2 + v^2 + t^2 = 1$ in the $s = 0$ hyper-plane. Let "E" denote the $L(S^2)$ ellipsoid image of S^2 (Figure 10.5). Then the linearity property of L implies, for $\alpha = (.4)$, for each point $\mathbf{p} = (u, v, 0, t)$ in S^2, and for $\mathbf{q} = (1/4, 1/4, 1/4, 1/4)$

(23) $L\left(.4S^2 + \mathbf{q}\right) = .4L(S^2) + L(\mathbf{q}) = .4E + (5/12, 5/12, 5/12).$

The argument "$.4S^2 + \mathbf{q}$" of L is the $z = 1/4$ great 2-sphere: Each point $(u, v, 0, t)$ of "S^2" scaled by "$(.4)$" corresponds to $(.4u, .4v, 0, .4t)$, and in turn, when translated by $(.25, .25, .25, .25)$, corresponds to

$$(x, y, 1/4, w) = (.4u + .25, .4v + .25, 0 + .25, .4t + .25)$$

which satisfies (12). Moreover, one may reverse the logic and show that each point $(x, y, 1/4, w)$ on the $w = 1/4$ great 2-sphere has such a *pre-image* point $(u, v, 0, t)$.

Turning to the L-image ".$4E + L(\mathbf{q})$" of "$(.4S^2 + \mathbf{q})$", we begin with the ellipsoid E pictured in Figure 10.5 and defined in (22). In equation (22) the ellipsoid E is defined in the context of the *eigenvector system* (x', y', z'). So we shall stay in this (x', y', z') system. We begin by scaling E by .4, which produces a "smaller ellipsoid": In detail, let $(x_s, y_s, z_s) = (.4x', .4y', .4z')$. Then using (22), the scaled down (smaller) ellipsoid is specified by

$$(24) \qquad \frac{x_s^2}{(.4/\sqrt{\lambda_1})^2} + \frac{y_s^2}{(.4/\sqrt{\lambda_2})^2} + \frac{z_s^2}{(.4/\sqrt{\lambda_3})^2} = 1$$

which references the (x', y', z') eigenvector system.

For the translation by "$L(\mathbf{q})$" we consider equation (5), which yields

$$\mathbf{x} = L(\mathbf{q}) = (5/12, 5/12, 5/12),$$

where these components of $\mathbf{x}$ refer to the (x, y, z) system, *not* the (x', y', z') eigenvector system. For calculation of the components of the vector $\mathbf{x}$ within the (x', y', z') system, we use the $\mathbf{x}' = P^T \mathbf{x}$ equation.

$$\begin{bmatrix} a \\ b \\ c \end{bmatrix} = \begin{bmatrix} 0 \\ \frac{(5/12)[2+(1-\lambda_2)]}{\sqrt{2+(1-\lambda_2)^2}} \\ \frac{(5/12)[2+(1-\lambda_3)]}{\sqrt{2+(1-\lambda_3)^2}} \end{bmatrix} = \begin{bmatrix} 1/\sqrt{2} & -1/\sqrt{2} & 0 \\ \frac{1}{\sqrt{2+(1-\lambda_2)^2}} & \frac{1}{\sqrt{2+(1-\lambda_2)^2}} & \frac{(1-\lambda_2)}{\sqrt{2+(1-\lambda_2)^2}} \\ \frac{1}{\sqrt{2+(1-\lambda_3)^2}} & \frac{1}{\sqrt{2+(1-\lambda_3)^2}} & \frac{(1-\lambda_3)}{\sqrt{2+(1-\lambda_3)^2}} \end{bmatrix} \begin{bmatrix} 5/12 \\ 5/12 \\ 5/12 \end{bmatrix}$$

So finally, within the (x', y', z') system with x' as "$x_s + a$", y' as "$y_s + b$", and z' as "$z_s + c$" we see that (24) implies

$$(25) \qquad \frac{(x'-a)^2}{(.4/\sqrt{\lambda_1})^2} + \frac{(y'-b)}{(.4/\sqrt{\lambda_2})^2} + \frac{(z'-c)}{(.4/\sqrt{\lambda_3})^2} = 1.$$

Equation (25) shows that the L image of "$.4S^2 + \mathbf{q}$" is an ellipsoid with center $(5/12, 5/12, 5/12)$ in the (x, y, z) frame.

§52 THE $y = 1/4$ AND $x = 1/4$ GREAT 2-SPHERES

The previous section concerned the "$z = 1/4$- *case*" — we obtained pictures of the $z = 1/4$ great 2-sphere and its L-image ellipsoid. This section concerns the "*x-case*" and "*y-case*", which amounts, respectively, to interchanging the first and third components in the z-case eigenvectors, and interchanging the second and third components in the z-case eigenvectors.

§52 THE y = 1/4 and x = 1/4 GREAT 2-SPHERES

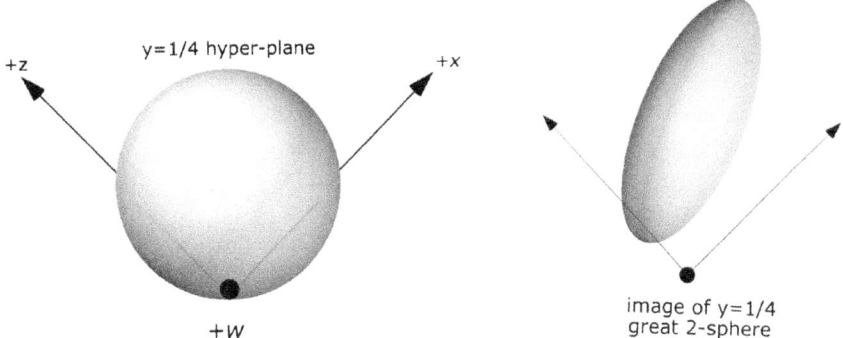

FIG. 10.6 The left-side scene is from inside the $x = 1/4$ hyper-plane. The scene contains the view of the $x = 1/4$ great 2-sphere from a camera located at $(1/4, 0, 0, 5)$ with target $(1/4, 1/3, 1/3, 1/3)$. The right-side scene is the (x, y, z) image space. The scene contains the view of the L-image ellipsoid from a camera located at $(0, 0, 5)$ with target $(1/3, 1/3, 1/3)$.

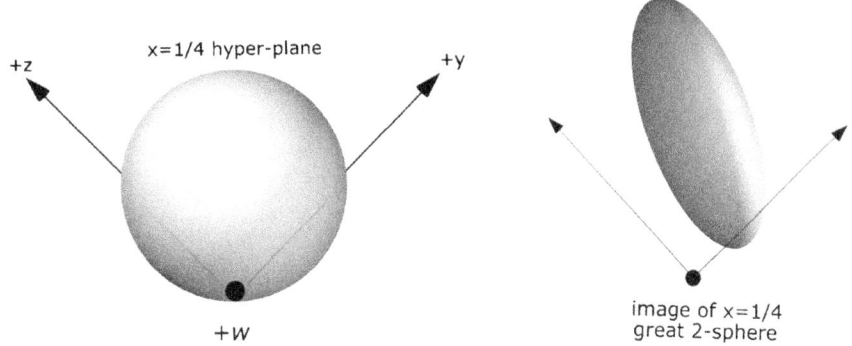

FIG. 10.7 The left-side scene is from inside the $y = 1/4$ hyper-plane. The scene contains the view of the $y = 1/4$ great 2-sphere from a camera located at $(0, 1/4, 0, 5)$ with target $(1/3, 1/4, 1/3, 1/3)$. The right-side scene is the (x, y, z) image space. The scene contains the view of the L-image ellipsoid from a camera located at $(0, 0, 5)$ with target $(1/3, 1/3, 1/3)$.

The development of the graphics for the x- and y-cases in Figures 10.7 and 10.6 run parallel to the z-case. To see the pattern, begin with the unit 3-sphere and three of its great 2-spheres:

$$u^2 + v^2 + s^2 + t^2 = 1$$

x-case	z-case	y-case
$(0, v, s, t)^T$	$(u, v, 0, t)^T$	$(u, 0, s, t)^T$

Then apply the linear transformation $L\left([u,v,s,t]^T\right) = [x,y,z]^T$, and solve for u, v, s, and t in terms of x, y, and z.

x-case	z-case	y-case
$v = y - x$	$u = x - z$	$u = x - y$
$s = z - x$	$v = y - z$	$s = z - y$
$t = \frac{3}{2}x$	$t = \frac{3}{2}z$	$t = \frac{3}{2}y$

Next, using $u^2 + v^2 + s^2 + t^2 = 1$ obtain quadratic forms and corresponding matrices M_x, M_z, and M_y.

x-case	z-case	y-case
$(y-x)^2 + (z-x)^2 + (\frac{3}{2}x)^2$	$(x-z)^2 + (y-z)^2 + (\frac{3}{2}z)^2$	$(x-y)^2 + (z-y)^2 + (\frac{3}{2}y)^2$

interchange z and x interchange z and y

$$M_x = \begin{bmatrix} \frac{17}{4} & -1 & -1 \\ -1 & 1 & 0 \\ -1 & 0 & 1 \end{bmatrix} \quad M_z = \begin{bmatrix} 1 & 0 & -1 \\ 0 & 1 & -1 \\ -1 & -1 & \frac{17}{4} \end{bmatrix} \quad M_y = \begin{bmatrix} 1 & -1 & 0 \\ -1 & \frac{17}{4} & -1 \\ 0 & -1 & 1 \end{bmatrix}$$

In each case, calculate the characteristic equation and eigenvalues.

x-case	z-case	y-case
same	$\leftarrow (\lambda - 1)\left[(\lambda - 1)(\lambda - \frac{17}{4}) - 2\right] = 0 \rightarrow$	same
same	$\leftarrow \lambda_1 = 1, \lambda_2 = \frac{21+3\sqrt{33}}{8}, \lambda_3 = \frac{21-3\sqrt{33}}{8} \rightarrow$	same

Then for M denoting the appropriate case, i.e., M_x, M_y, or M_z, use the equation $M\left([x,y,z]^T\right) = \lambda\left([x,y,z]^T\right)$ to obtain, in each case, three equations in three unknowns.

x-case	z-case	y-case
$\frac{17}{4}x - y - z = \lambda x$	$x + 0 - z = \lambda x$	$x - y + 0 = \lambda x$
$-x + y + 0 = \lambda y$	$0 + y - z = \lambda y$	$-x + \frac{17}{4}y - z = \lambda y$
$-x + 0 + z = \lambda z$	$-x - y + \frac{17}{4}z = \lambda z$	$0 - y + z = \lambda z$

interchange z and x interchange z and y

The solutions yield eigenvectors:

x-case:
$$\begin{bmatrix} 1-\lambda_3 \\ 1 \\ 1 \end{bmatrix} \begin{bmatrix} 1-\lambda_2 \\ 1 \\ 1 \end{bmatrix} \begin{bmatrix} 0 \\ -1 \\ 1 \end{bmatrix}$$

z-case:
$$\begin{bmatrix} 1 \\ -1 \\ 0 \end{bmatrix} \begin{bmatrix} 1 \\ 1 \\ 1-\lambda_2 \end{bmatrix} \begin{bmatrix} 1 \\ 1 \\ 1-\lambda_3 \end{bmatrix}$$

y-case:
$$\begin{bmatrix} 1 \\ 0 \\ -1 \end{bmatrix} \begin{bmatrix} 1 \\ 1-\lambda_3 \\ 1 \end{bmatrix} \begin{bmatrix} 1 \\ 1-\lambda_2 \\ 1 \end{bmatrix}$$

After normalizing the *x-case* eigenvectors, we let

$$P_{x\text{-case}} = \begin{bmatrix} \frac{1-\lambda_3}{\sqrt{2+(1-\lambda_3)^2}} & \frac{1-\lambda_2}{\sqrt{2+(1-\lambda_2)^2}} & 0 \\ \frac{1}{\sqrt{2+(1-\lambda_3)^2}} & \frac{1}{\sqrt{2+(1-\lambda_2)^2}} & \frac{-1}{\sqrt{2}} \\ \frac{1}{\sqrt{2+(1-\lambda_3)^2}} & \frac{1}{\sqrt{2+(1-\lambda_2)^2}} & \frac{1}{\sqrt{2}} \end{bmatrix}.$$

Likewise, normalizing the *y-case* eigenvectors, we let

$$P_{y\text{-case}} = \begin{bmatrix} \frac{1}{\sqrt{2}} & \frac{1}{\sqrt{2+(1-\lambda_3)^2}} & \frac{1}{\sqrt{2+(1-\lambda_2)^2}} \\ 0 & \frac{1-\lambda_3}{\sqrt{2+(1-\lambda_3)^2}} & \frac{1-\lambda_2}{\sqrt{2+(1-\lambda_2)^2}} \\ \frac{-1}{\sqrt{2}} & \frac{1}{\sqrt{2+(1-\lambda_3)^2}} & \frac{1}{\sqrt{2+(1-\lambda_2)^2}} \end{bmatrix}.$$

These matrices coupled with the M_x and M_y matrices yield canonical forms of ellipsoids. In particular, as detailed in the *z-case* equation (22), the $P_x = P_{x\text{-case}}$ and $P_y = P_{y\text{-case}}$ matrices yield diagonal matrices according to

$$P_x^T M_x P_x = \begin{bmatrix} \lambda_3 & 0 & 0 \\ 0 & \lambda_2 & 0 \\ 0 & 0 & \lambda_1 \end{bmatrix} \quad \text{and} \quad P_y^T M_y P_y = \begin{bmatrix} \lambda_1 & 0 & 0 \\ 0 & \lambda_3 & 0 \\ 0 & 0 & \lambda_2 \end{bmatrix}.$$

Letting $\mathbf{e}'_1$, $\mathbf{e}'_2$, and $\mathbf{e}'_3$ denote respectively the first, second, and third column vectors in P_x, we define the $+x'$, $+y'$, and $+z'$ directions that provide the (x', y', z') representation of the quadratic form in the *x-case*

$$z^2 + y^2 + (17/4)x^2 - 2zx - 2zy = \lambda_3(x')^2 + \lambda_2(y')^2 + \lambda_1(z')^2.$$

And letting $\mathbf{e}''_1$, $\mathbf{e}''_2$, and $\mathbf{e}''_3$ denote respectively the first, second, and third column vectors in P_y, we define $+x''$, $+y''$, and $+z''$ directions that provide the (x'', y'', z'') representation of the quadratic form in the *y-case*

$$x^2 + z^2 + (17/4)y^2 - 2xy - 2xz = \lambda_1(x'')^2 + \lambda_3(y'')^2 + \lambda_2(z'')^2.$$

The context for the *x-* and *y-cases* rests within the *z-case* eigenvectors $\mathbf{e}_1$, $\mathbf{e}_2$, and $\mathbf{e}_3$ introduced above (22). We now denote $\mathbf{e}_1$, $\mathbf{e}_2$, and $\mathbf{e}_3$ as, respectively, $\mathbf{e}'''_1$, $\mathbf{e}'''_2$, and $\mathbf{e}'''_3$. With this notational change, we provide meaning to the directions $+x'''$, $+y'''$, and $+z'''$. Since each of the *x-y-z*-forms equate to the value 1, we have equations analogous to (22)

$$\begin{aligned} x\text{-case}: & \quad \frac{x'^2}{\left(1/\sqrt{\lambda_3}\right)^2} + \frac{y'^2}{\left(1/\sqrt{\lambda_2}\right)^2} + \frac{z'^2}{\left(1/\sqrt{\lambda_1}\right)^2} = 1 \\ y\text{-case}: & \quad \frac{x''^2}{\left(1/\sqrt{\lambda_1}\right)^2} + \frac{y''^2}{\left(1/\sqrt{\lambda_3}\right)^2} + \frac{z''^2}{\left(1/\sqrt{\lambda_2}\right)^2} = 1 \\ z\text{-case}: & \quad \frac{x'''^2}{\left(1/\sqrt{\lambda_1}\right)^2} + \frac{y'''^2}{\left(1/\sqrt{\lambda_2}\right)^2} + \frac{z'''^2}{\left(1/\sqrt{\lambda_3}\right)^2} = 1 \end{aligned}$$

Keeping in mind that these equations represent three ellipsoids centered at the origin of the (x, y, z) system, we must, as we did in the z-case following (22), scale by "(.4)" and then translate to the point $\mathbf{x} = L(\mathbf{q})$ where $\mathbf{q} = (.25, .25, .25, .25)$. In each of the frames we calculate

$$\begin{array}{ccc} x\text{-case} & z\text{-case} & y\text{-case} \\ [a', b', c']^T = P_x^T \mathbf{x} & [a''', b''', c''']^T = P_z^T \mathbf{x} & [a'', b'', c'']^T = P_y^T \mathbf{x} \end{array}$$

With these three representations — the primed, double primed, and triple primed — of the common center of the three ellipsoids, we may express the three desired canonical forms

$$x\text{-case}: \quad \frac{(x'-a')^2}{\left(.4/\sqrt{\lambda_3}\right)^2} + \frac{(y'-b')^2}{\left(.4/\sqrt{\lambda_2}\right)^2} + \frac{(z'-c')^2}{\left(.4/\sqrt{\lambda_1}\right)^2} = 1$$

$$y\text{-case}: \quad \frac{(x''-a'')^2}{\left(.4/\sqrt{\lambda_1}\right)^2} + \frac{(y''-b'')^2}{\left(.4/\sqrt{\lambda_3}\right)^2} + \frac{(z''-c'')^2}{\left(.4/\sqrt{\lambda_2}\right)^2} = 1$$

$$z\text{-case}: \quad \frac{(x'''-a''')^2}{\left(.4/\sqrt{\lambda_1}\right)^2} + \frac{(y'''-b''')^2}{\left(.4/\sqrt{\lambda_2}\right)^2} + \frac{(z'''-c''')^2}{\left(.4/\sqrt{\lambda_3}\right)^2} = 1$$

§53 Comments

The motivation for writing this Chapter originated from the author's reading of Part III Chapter 14 *The Hypersphere* in Jeffrey Weeks' book [38]. Weeks' Figures 14.4 and 14.5 were especially interesting — Figure 14.5 explains how to "think about" four great 2-spheres in a hyper-sphere.

The upshot of this chapter is that the transformation L that we used to create the "God's Image?" graphic transforms each of the four great 2-spheres specified in this chapter into the shape of an ellipsoid — one of which is a sphere. The goal of the next chapter is to locate the positions of these ellipsoids within the "God's Image?" graphic. *Will we be able to see an ellipsoid of points within the graphic?*

In passing, we note that like the "God's Image?" graphic, Poincaré would agree that the ellipsoidal images *were not interchanged and their relative situation was conserved.*

CHAPTER 11

Images of Great 2-spheres

The hyper-sphere "$(x-1/4)^2+(y-1/4)^2+(z-1/4)^2+(w-1/4)^2=(.4)^2$" and its four great 2-spheres obtained by respectively setting $x = 1/4$, $y = 1/4$, $z = 1/4$, and $w = 1/4$ were studied extensively in Chapter 10. In particular, equations for their entire L images in 3-space were derived. Because these great 2-sphere are contained within the hyper-sphere, they invite the question *Can we see partial images of these 2-spheres inside the* God's Image? *graphic?* This chapter, supplemented with corresponding color plates, provides an answer.

The *L image* appears in an "(x', y', z')" system — as presented below, the x' direction points up and to the right, the y' points up and to the left, and the z' comes out of the paper toward the reader.

§54 THE $w = 1/4$ CASE

The "$w = 1/4$ case" concerns the L image of the great 2-sphere specified by those points (x, y, z, w) in the hyper-sphere whose w component is fixed with value 1/4.

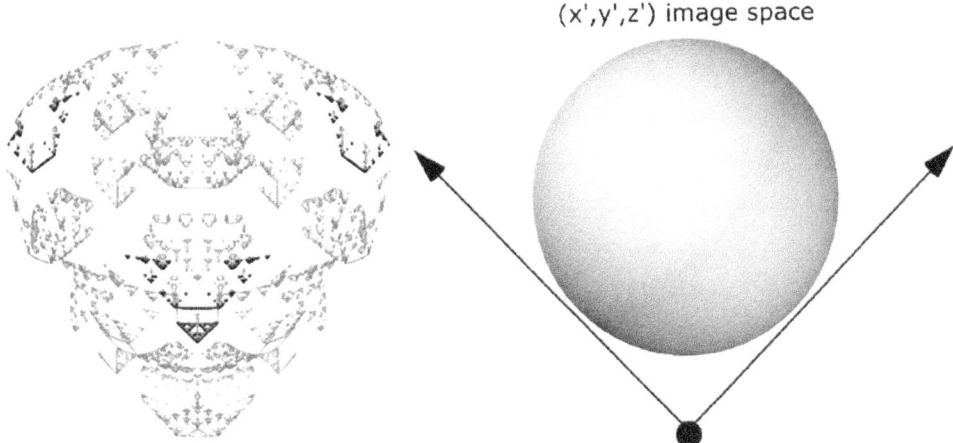

FIG. 11.1 The entire L image of the $w = 1/4$ great 2-sphere is pictured on the right; on the left we see its partial image (black dots) contained within the "God's Image?" graphic. Color plate? The black dots appear as green dots.

§55 THE $z = 1/4$ CASE

The "$z = 1/4$ case" concerns the L image of the great 2-sphere specified by those points (x, y, z, w) in the hyper-sphere whose z component is fixed with value $1/4$.

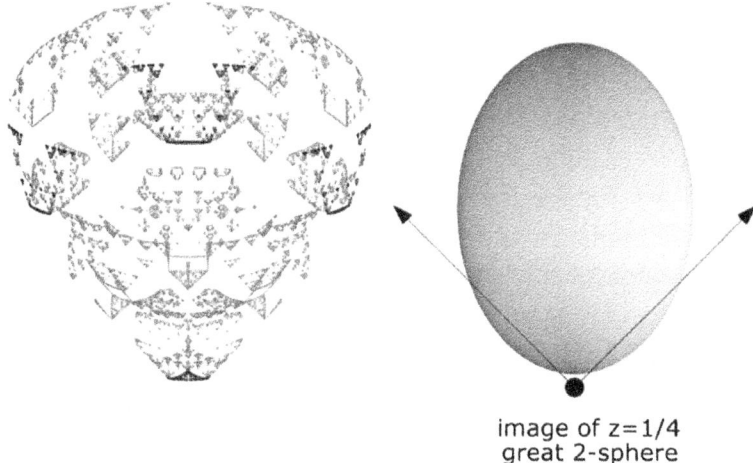

FIG. 11.2 The entire L image of the $z = 1/4$ great 2-sphere is pictured on the right; on the left we see its partial image (black dots) contained within the "God's Image?" graphic. Color plate? The black dots appear as various colored dots.

§56 THE $y = 1/4$ CASE

The "$y = 1/4$ case" concerns the L image of the great 2-sphere specified by those points (x, y, z, w) in the hyper-sphere whose y component equals $1/4$.

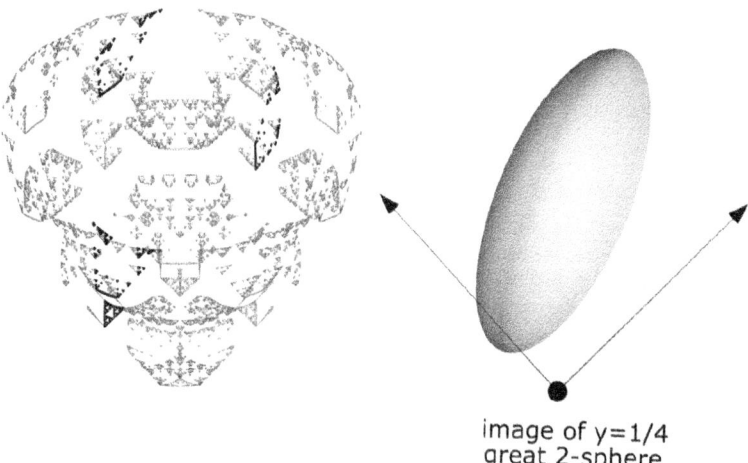

FIG. 11.3 The entire L image of the $y = 1/4$ great 2-sphere is pictured on the right; on the left we see its partial image (black dots) contained within the "God's Image?" graphic. Color plate? The black dots appear as various colored dots.

§57 The $x = 1/4$ case

The "$x = 1/4$ case" concerns the L image of the great 2-sphere specified by those points (x, y, z, w) in the hyper-sphere whose x component equals $1/4$.

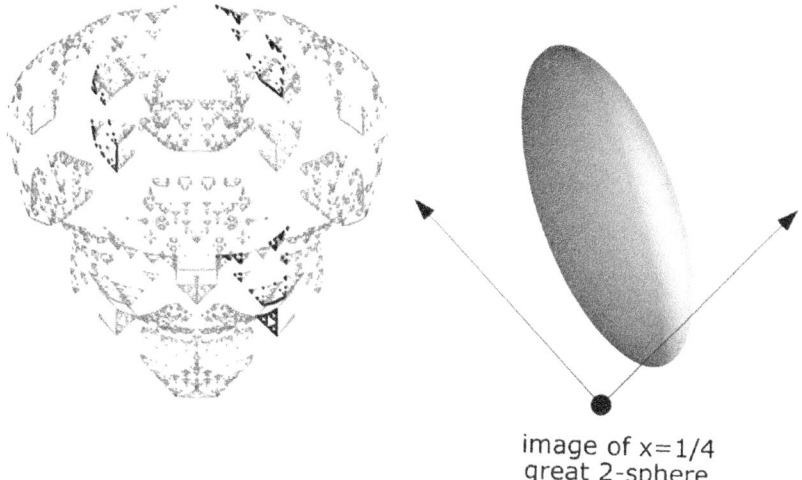

image of x=1/4
great 2-sphere

FIG. 11.4 The entire L image of the $x = 1/4$ great 2-sphere is pictured on the right; on the left we see its partial image (black dots) contained within the "God's Image?" graphic. Color plate? The black dots appear as various colored dots.

§58 Topological images

The art pictured in Figures 11.1 through 11.4 motivates more art. In particular, it is natural to bundle (join) various cases in order to approach the global structure. Images of the "joins" of the various cases are provided in §59; the question about these "joins" being topological representations of their 4-space pre-images are motivated in §60 and then discussed in §61.

In this section we focus on the *God's Image?* graphics and great 2-sphere images in Figures 11.1 through 11.4. We find that each of the graphics images in these figures are *topological representations* of parts of objects beyond human view.

But what is a topological representation? The idea is explained in any book on topology, but here we shall recall the words that Poincaré's used to express his idea of a "topological representation" (see Preface)

> ... geometry is the art of reasoning well with badly made figures. Yes, without doubt, but with one condition. The proportions of the figures might be grossly altered but their elements must not be interchanged and must conserve their relative situation.

For those interested in more detail, a mathematical proof, expressed in modern terms, that the *God's Image?* graphic is a topological representation of those points in 4-dimensional space that are common to a 3-sphere and a 4-web grid may be viewed as a corollary to Theorem 67.3 on page 138 of [18]. And from §49 in this book, the argument following equation (11) shows that the L images — the one 2-sphere and 3 ellipsoids illustrated in Figures 11.1 through 11.4 — are topological representations of the great 2-spheres (10).

§59 IMAGES OF JOINS

We begin by considering the join of the $x = 1/4$ and $y = 1/4$ great 2-spheres.

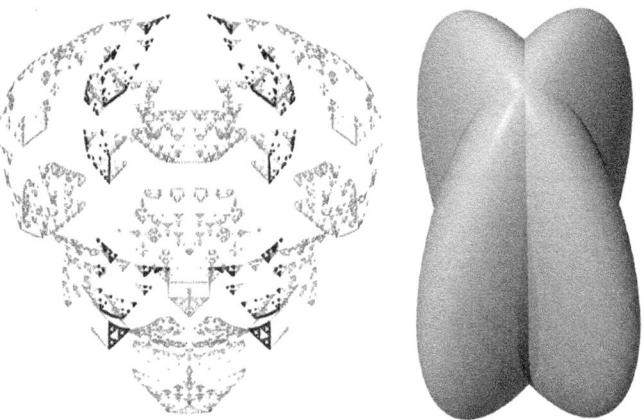

FIG. 11.5 The entire and partial image of the join of $x = y = 1/4$ great 2-spheres.

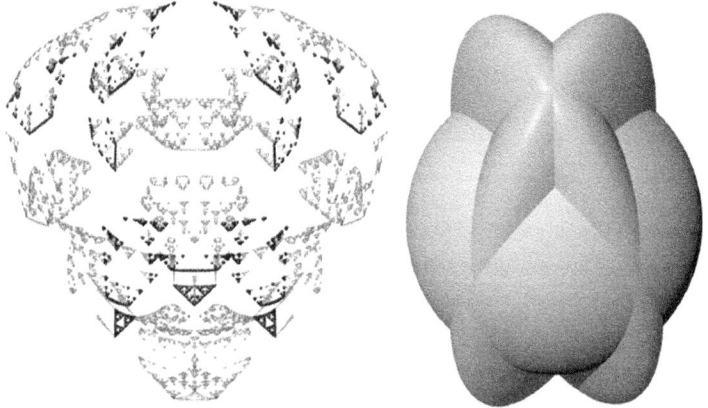

FIG. 11.6 Entire and partial image of join of $x = y = w = 1/4$ great 2-spheres.

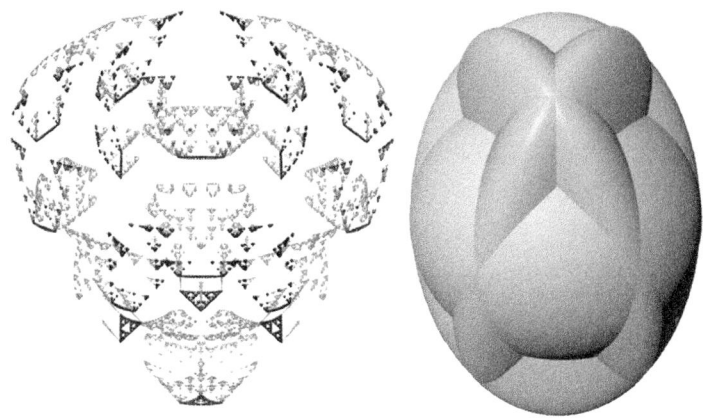

FIG. 11.7 Entire and partial image of join of $x = y = z = w = 1/4$ great 2-spheres.

But what about the graphics in Figures 11.5, 11.6, and 11.7 — are those graphics topological representations?

§60 LOWER-DIMENSIONAL EXAMPLE

To aid our understanding of *topological representation* — Poincaré's *proportions of the figures might be grossly altered but their elements must not be interchanged and must conserve their relative situation* — let's restrict our transformation L to the unit 2-sphere, namely,

$$(1) \qquad u^2 + v^2 + s^2 = 1 \quad \text{and} \quad \begin{bmatrix} x \\ y \end{bmatrix} = \begin{bmatrix} 1 & 0 & 2/3 \\ 0 & 1 & 2/3 \end{bmatrix} \begin{bmatrix} u \\ v \\ s \end{bmatrix}.$$

Define the great circles C_v and C_s whose points, respectively, have their v component and s component equal to zero. That is,

$$C_v: \quad u^2 + 0^2 + s^2 = 1 \quad \text{and} \quad \begin{bmatrix} u + (2/3)s \\ (2/3)s \end{bmatrix} = \begin{bmatrix} 1 & 0 & 2/3 \\ 0 & 1 & 2/3 \end{bmatrix} \begin{bmatrix} u \\ 0 \\ s \end{bmatrix};$$

$$C_s: \quad u^2 + v^2 + 0^2 = 1 \quad \text{and} \quad \begin{bmatrix} u \\ v \end{bmatrix} = \begin{bmatrix} 1 & 0 & 2/3 \\ 0 & 1 & 2/3 \end{bmatrix} \begin{bmatrix} u \\ v \\ 0 \end{bmatrix}.$$

Note that L fixes C_s, i.e., "$L(C_s) = C_s$". But in the C_v case L transforms the circle C_v into an ellipse. Indeed, consider the following:

$$\left. \begin{array}{l} x = u + (2/3)s \\ y = (2/3)s \end{array} \right\} \iff \left\{ \begin{array}{l} u = x - y \\ s = (3/2)y \end{array} \right.$$

which, when coupled with $u^2 + s^2 = 1$, provides

(2) $$(x - y)^2 + ((3/2)y)^2 = 1.$$

This equation (2) is equation (8) in Chapter 10. Moreover, because $v = 0$, our restricted transformation L may be represented as

(3) $$u^2 + s^2 = 1 \quad \text{and} \quad \begin{bmatrix} x \\ y \end{bmatrix} = \begin{bmatrix} 1 & 2/3 \\ 0 & 2/3 \end{bmatrix} \begin{bmatrix} u \\ s \end{bmatrix},$$

which shows that the calculations in §48 expose the L-image of C_v as an ellipse — the transformation in (3) above also appears in equation (6) of Chapter 10; and at the end of section §48 we find the canonical form of *our restricted L-image* of C_v:

$$\frac{x'^2}{a^2} + \frac{y'^2}{b^2} = 1$$

where $a = 1/\sqrt{\lambda_1} > 0$ and $b = 1/\sqrt{\lambda_2} > 0$ for eigenvalues

$$\lambda_1 = .619800677654\ldots \quad \text{and} \quad \lambda_2 = 3.63019932236\ldots$$

with unit eigenvectors

$$\mathbf{e}_1 = \begin{bmatrix} \cos\theta \\ \sin\theta \end{bmatrix} \quad \text{and} \quad \mathbf{e}_2 = \begin{bmatrix} -\sin\theta \\ \cos\theta \end{bmatrix} \quad \text{for} \quad \theta = 20.816769\ldots \text{ degrees.}$$

These eigenvectors determine, respectively, positive directions of the x' axis and the y' axis. The ellipse has its major axis of length $2a$ and its minor axis length $2b$.

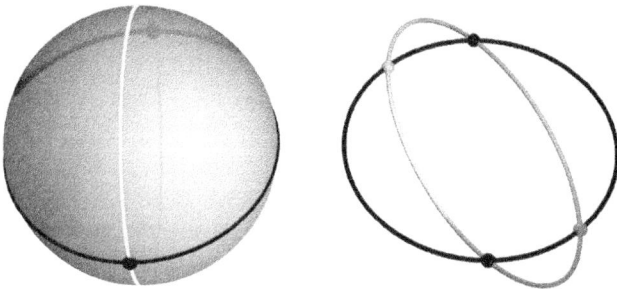

Graphically speaking, we see only *two antipodal points* (black dots on the left) being shared by circles C_s and C_v while their images (black circle and gray ellipse on the right) share *four points* — two black and two gray dots.

On inspection of this graphic Poincaré might say, "The transformation of these circles and their shared points *does not* conserve their relative situation". In modern terms, the transformation is not topological and the image is therefore not a topological image of two circles meeting at *only* two points.

Within the graphic above, the two black antipodal dots on the left are specified by $[u, v, s]^T = [1, 0, 0]^T$ and $[-1, 0, 0]^T$. Also, $L([1, 0, 0]^T) = [1, 0]^T$ and $L([-1, 0, 0]^T) = [-1, 0]^T$ show that the two black dots on the right are likewise antipodal. The obvious question arises, *Where are the pre-images of the gray dots?* To answer this question, we first view our restriction of L as a group homomorphism whose kernel is easily calculated as

$$L([u, v, s]^T) = [0, 0]^T \Leftrightarrow \begin{bmatrix} u \\ v \\ s \end{bmatrix} = \begin{bmatrix} -(2/3)q \\ -(2/3)q \\ q \end{bmatrix} \quad \text{for some scalar ``}q\text{''.}$$

If $q = 0$, then $[u, v, s]^T = [0, 0, 0]^T$ is neither a point in C_s nor a point in C_v. So we shall suppose $q \neq 0$. It turns out that distinct $[u, v, 0]^T$ in C_s and $[u', 0, s']^T$ in C_v are transformed by L to a single point $L([u, v, 0]^T) = L([u', 0, s']^T)$ whenever

$$(4) \qquad \begin{bmatrix} u \\ v \\ 0 \end{bmatrix} - \begin{bmatrix} u' \\ 0 \\ s' \end{bmatrix} = \begin{bmatrix} -(2/3)q \\ -(2/3)q \\ q \end{bmatrix}.$$

Since $u^2 + v^2 = 1$ and $(u')^2 + (s')^2 = 1$ there are angles θ and ψ such that

$$\begin{aligned} u - u' &= \cos\theta - \cos\psi = -(2/3)q \\ v - 0 &= \sin\theta = -(2/3)q \\ 0 - s' &= -\sin\psi = q. \end{aligned}$$

So "$u - u' = v$", or "$\cos\theta - \cos\psi = \sin\theta$". After solving for $\cos\psi$, we substitute into $\cos^2\psi + \sin^2\psi = 1$ to obtain

$$(\cos\theta - \sin\theta)^2 + \sin^2\psi = 1 \quad \text{which yields} \quad \sin^2\psi = 2\cos\theta\sin\theta.$$

Then since $-s' = -\sin\psi = q$ and $\sin\theta = -(2/3)q$, we have $(-q)^2 = -2\cos\theta(2/3)q$, or, $\cos\theta = -(3/4)q$. And again using $\sin\theta = -(2/3)q$, we may substitute into $\sin^2\theta + \cos^2\theta = 1$ and solve

$$(-(2/3)q)^2 + (-(3/4)q)^2 = 1, \quad \text{or,} \quad q = 12/(\pm\sqrt{145}).$$

For each of the two values of $q = 12/(\pm\sqrt{145})$ the equations $u^2 + v^2 = 1$ and $u - u' = -(2/3)q$ yield corresponding values for u, v, u', and s'. We thereby obtain two solutions to (4) that yield pairs of antipodal points. The solutions, for $d = +\sqrt{145}$, are

$$\begin{bmatrix} 9/d \\ 8/d \\ 0 \end{bmatrix} - \begin{bmatrix} 1/d \\ 0 \\ 12/d \end{bmatrix} = \begin{bmatrix} 8/d \\ 8/d \\ -12/d \end{bmatrix} \quad \text{and} \quad \begin{bmatrix} -9/d \\ -8/d \\ 0 \end{bmatrix} - \begin{bmatrix} -1/d \\ 0 \\ -12/d \end{bmatrix} = \begin{bmatrix} -8/d \\ -8/d \\ 12/d \end{bmatrix}.$$

And these matrix equations are illustrated below, where the parallel *straight-shaft* arrows going in opposite directions represent the vectors on the right-side of the equalities. The light-gray antipodal dots appear on the light-gray circle C_v, and their L-images appear as dark-gray antipodal dots on the black circle C_s. The arrows from the light-gray to dark-gray dots serve to demonstrate the action of L on the light-gray dots. Similarly the black *circle-arrows* that start and end at the dark-gray dots demonstrate the action of L *fixing* the dark-gray dots on the black circle C_s. So together these four arrows show that each of the two dark-gray dots have two pre-images — one straight- and one circle-arrow.

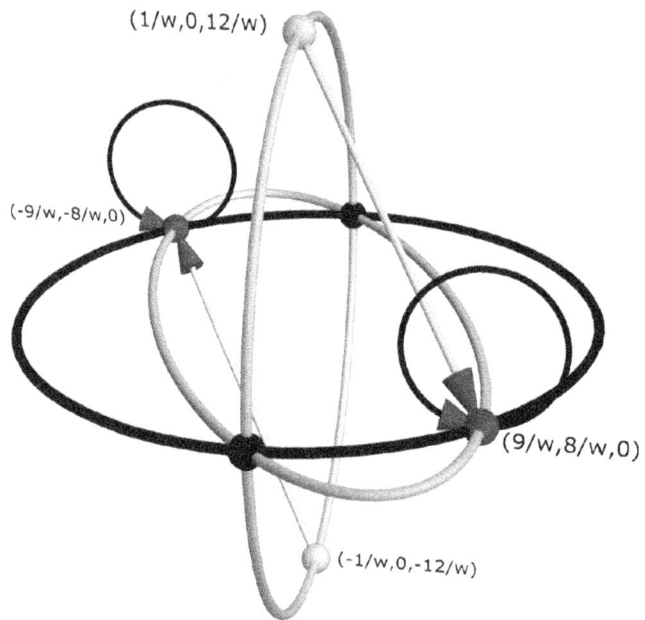

§61 Joins of Great 2-spheres

In this section we consider the question *Are the L images represented by the right-side graphics within the figures of §59 topological representations?* The answer is simply that *none* of the right-side graphics within the figures of §59 represent topological representations.

To see why this is the case, we shall show that the L image of the hyper-sphere

$$(x - 1/4)^2 + (y - 1/4)^2 + (z - 1/4)^2 + (w - 1/4)^2 = (.4)^2$$

restricted to the join (union) of *any two* of our great 2-sphere slices — slice $x = 1/4$, slice $y = 1/4$, slice $z = 1/4$, or slice $w = 1/4$ — fails to be a topological representation. The proof rests on the fact that each such join contains at least one point that represents the image of two distinct points from the pre-image join in 4-space.

The argument is greatly simplified by the fact that scaling and translation within 3-dimensional space does not change the topological representation. For example, if you shrink a sphere or an ellipsoid and then move it the result is a topological representation of the original. The sphere example is illustrated in Figure 10.2.

Our focus is the *unit* hyper-sphere in 4-dimensional space:

$$x^2 + y^2 + z^2 + w^2 = 1.$$

Following the lower-dimensional example presented in the previous section, we introduce similar notation, i.e., S_t denotes $S_{t=0}$, and, $S_v = S_{v=0}$.

$$S_v : u^2 + 0^2 + s^2 + t^2 = 1, \text{ and, } \begin{bmatrix} u + (2/3)t \\ (2/3)t \\ s + (2/3)t \end{bmatrix} = \begin{bmatrix} 1 & 0 & 0 & 2/3 \\ 0 & 1 & 0 & 2/3 \\ 0 & 0 & 1 & 2/3 \end{bmatrix} \begin{bmatrix} u \\ 0 \\ s \\ t \end{bmatrix};$$

$$S_t : u^2 + v^2 + s^2 + 0^2 = 1, \text{ and, } \begin{bmatrix} u \\ v \\ s \end{bmatrix} = \begin{bmatrix} 1 & 0 & 0 & 2/3 \\ 0 & 1 & 0 & 2/3 \\ 0 & 0 & 1 & 2/3 \end{bmatrix} \begin{bmatrix} u \\ v \\ s \\ 0 \end{bmatrix}.$$

Note that L fixes S_t, i.e., "$L(S_t) = S_t$". But in the S_v case, $L(S_v)$ is an ellipsoid whose scaled and translated graphic appears on the right-side in Figure 11.2. Our goal is to show that $L(S_v)$ joined with $L(S_t)$ in 3-space is not a topological representation of S_v joined with S_t in 4-space.

As for the transformation L, we begin by viewing L as a group homomorphism whose kernel is easily calculated.

$$L([u,v,s,t]^T) = [0,0,0]^T \Leftrightarrow \begin{bmatrix} u \\ v \\ s \\ t \end{bmatrix} = \begin{bmatrix} -(2/3)q \\ -(2/3)q \\ -(2/3)q \\ q \end{bmatrix} \text{ for some scalar "}q\text{".}$$

It turns out that distinct $[u,0,s,t]^T$ in S_v and $[u,v,s,0]^T$ in S_t are transformed by L to a single point $L([u,0,s,t]^T) = L([u,v,s,0]^T)$ whenever

$$\begin{bmatrix} u \\ 0 \\ s \\ t \end{bmatrix} - \begin{bmatrix} u' \\ v' \\ s' \\ 0 \end{bmatrix} = \begin{bmatrix} -(2/3)q \\ -(2/3)q \\ -(2/3)q \\ q \end{bmatrix}.$$

Finding a solution to this case is easy if we follow the example presented in the previous section. In particular, recall that the two values of q were

$q = (12/\pm\sqrt{145})$. The number "145" turns out to be the common value of two sums of squares, namely,

$$9^2 + 8^2 + 0^2 = 145 \quad \text{and} \quad 1^2 + 0^2 + 12^2 = 145.$$

which are, respectively, the numerators of the first and second vectors listed at the bottom of page 115. Thus, for $q = 12/(+\sqrt{145})$ and $d = (+\sqrt{145})$, we see that

$$(9/d)^2 + (8/d)^2 + (0/d)^2 = 1 \quad \text{and} \quad (1/d)^2 + (0/d)^2 + (12/d)^2 = 1$$

which shows that the first and second vectors listed at the bottom of page 115 are unit vectors. The key is that "8" is 2/3 of "12".

Analogously, for the S_v and S_t case at hand, consider $d = +\sqrt{161}$, $q = -12/d$, and

$$\begin{bmatrix} 9/d \\ 8/d \\ 4/d \\ 0 \end{bmatrix} - \begin{bmatrix} 1/d \\ 0 \\ -4/d \\ 12/d \end{bmatrix} = \begin{bmatrix} 8/d \\ 8/d \\ 8/d \\ -12/d \end{bmatrix}.$$

As for the other two pairs that include S_t, we may permute rows to obtain similar distinct points whose L image is a single point. In particular, if S_u is the sphere joined with S_t, then in the equation above permute the u and v rows in each of the left-side vectors.

For the three pairs of great 2-spheres that do not involve S_t, consider the example of the S_u join S_v. In this case, let $d = (+\sqrt{116})$ and $q = 12/d$. The corresponding equation is:

$$\begin{bmatrix} 0 \\ -8/d \\ -4/d \\ 6/d \end{bmatrix} - \begin{bmatrix} 8/d \\ 0 \\ 4/d \\ -6/d \end{bmatrix} = \begin{bmatrix} -8/d \\ -8/d \\ -8/d \\ 12/d \end{bmatrix}.$$

To verify that we are headed in the right direction in this S_u-join-S_v case, note the calculation below — the L image point $[4/d, -4/d, 0]$ has the two L pre-image points that appear above as the two distinct left-side vectors.

$$\begin{bmatrix} 4/d \\ -4/d \\ 0 \end{bmatrix} = \begin{bmatrix} 4/d \\ -8/d + 4/d \\ -4/d + 4/d \end{bmatrix} = \begin{bmatrix} 1 & 0 & 0 & 2/3 \\ 0 & 1 & 0 & 2/3 \\ 0 & 0 & 1 & 2/3 \end{bmatrix} \begin{bmatrix} 0 \\ -8/d \\ -4/d \\ 6/d \end{bmatrix}$$

and

$$\begin{bmatrix} 4/d \\ -4/d \\ 0 \end{bmatrix} = \begin{bmatrix} 8/d - 4/d \\ -4/d \\ 4/d - 4/d \end{bmatrix} = \begin{bmatrix} 1 & 0 & 0 & 2/3 \\ 0 & 1 & 0 & 2/3 \\ 0 & 0 & 1 & 2/3 \end{bmatrix} \begin{bmatrix} 8/d \\ 0 \\ 4/d \\ -6/d \end{bmatrix}$$

For the remaining two joins, namely, S_u-join-S_s, and, S_v-join-S_s, we may appropriately permute the rows of the S_u-join-S_v case, leaving the value of $d = (+\sqrt{116})$ unchanged.

§62 COMMENTS

Figures 11.1, 11.2, 11.3, and 11.4 contain two graphics each. And each graphic is a topological representation (topological image) of the corresponding structure from the fourth dimension. The movement from 4-space into 3-space is an application of the linear transformation L. And while the graphics on the left side of each of these figures may be described as *grossly altered* (Poincaré's words), *their elements ... are not interchanged and they conserve their relative situation* (Poincaré paraphrased). As to the graphics on the right side of these figures, it is easy see that each ellipsoid/2-sphere is a topological image of a 2-sphere — recall equation (11) of Chapter 10 and its consequences discussed thereafter.

As to Figures 11.5, 11.6, and 11.7, however, only the left-side graphic in each figure is a topological representation. The argument presented in §60 served to motivate the arguments in §61 where proofs were provided that the right-side graphics in these three figures are not topological representations.

Finally, recalling the reference to Jeffrey Weeks' book [38] in §53 above, especially the reference to his presentation of four great 2-spheres in his Figure 14.5, one could compare the right-side graphic of Figure 11.7 above with his Figure 14.5. Neither is a topological image of four great 2-spheres imbedded within a hyper-sphere, but the left-side graphic of Figure 11.7 above *is* a topological representation of a *partial image* of four great 2-spheres in a hyper-sphere.

A proof that the *partial image* (the graphic we call *God's Image?*) is a topological image of the "captured points" on our hyper-sphere follows from the following: *All* of the "captured points" reside inside our *4-space* 4-web, and, our 3-space 4-web is a topological image of our 4-space 4-web. (See author's Springer Book [20]).

CHAPTER 12

Appendix 1: Supplement for Chapters 1 and 2

This Appendix provides more rigor for those who may desire more in-depth information surrounding §4 *Gluing discs*, §5 *Slicing spheres*, and §10 *Dante's 3-sphere*. With regard to §4, we provide §A1 *Flexing a disc* and §A2 *3-disc flex step*. For §5 we provide the mathematics underlying the slicings of a 3-sphere — see §A5 *Slices of a 3-sphere* and §A6 *Hyperplane slices of the 3-sphere*. Both §A5 and §A6 require some technical background.

For Chapter 1 in general, there is §A4 *Patterns and the 3-sphere*, which is a concise summary of the key concepts. And for Chapter 2 §10 *Dante's 3-sphere*, we provide §A3 *Using a mirror to glue discs* for insight into how Dante's description of a "reflection in a mirror" equates to the gluing process.

§A1 FLEXING A DISC

We begin with the "flex step" in the 1-disc case. In Figure A1.1 we see the *horizontal line* ℓ and the *vertical line* ℓ'. These lines define two of the three dimensions that we can see. To place glue on the points p and q of the 1-disc $[p,q]$, we need to keep p and q within the line ℓ (the horizontal dimension) while we use the vertical dimension to move the *inside of* $[p,q]$ off of ℓ.[1]

Observe that we can simultaneously see both the inside of the "flexed 1-disc" (semicircle) and the S^0 boundary because our construction involved only two of our three "visual dimensions."

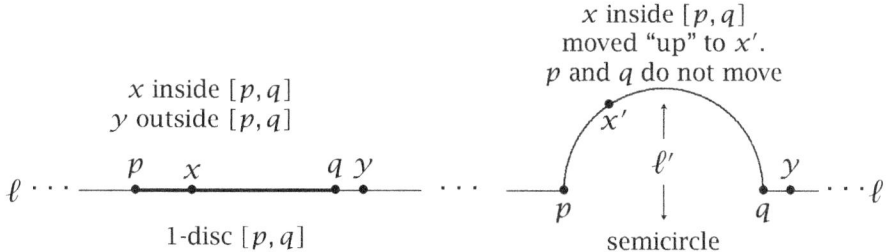

FIG. A1.1 The inside of a 1-disc is moved into another dimension.

---

[1]Think about it, if you are going to glue two objects together at specified locations, you must first position the objects so that whenever you apply the glue, only the specified locations are glued.

Even though Figure A1.3 below shows us that 2-dimensional vision allows us to see a "flexed 1-disc" (semicircle), could it be that we can also see a semicircle if we had only 1-dimensional vision?

Perhaps Freddy the penguin with one-dimensional vision can illustrate the answer to this question (Figure A1.2).

FIG. A1.2 Freddy with one-dimensional vision cannot see a 1-disc.

So it appears that we need at least 2-dimensional vision to see a "flexed 1-disc" (semicircle). To be sure, let's give Freddy 2-dimensional vision, and look at what he sees (Figure A1.3):

FIG. A1.3 Freddy sees the entire 1-disc and a "flexed 1-disc" (semicircle).

As for the "flex step" in the 2-disc case, we need our third dimension of vision. That is, first picture a 2-disc on a sheet of "paper" containing a horizonal line ℓ and a vertical line ℓ'. The "paper" accounts for two of our three "visual dimensions." Then we introduce a third line ℓ''' that injects a third dimension for moving the inside of our 2-disc off the "paper" while the S^1 boundary remains within the "paper".[2]

---

[2] We may picture the "paper" that contains the 2-disc as lying flat on a table top, and then picture the line ℓ''' as a vertical line going through the paper toward the ceiling.

Since our construction of the "flexed 2-disc" (hemisphere) requires only 3-dimensions, we can, as is illustrated in Figure 1.6 of Chapter 1, simultaneously picture both the "hemispheric 2-disc" and the bounding one-sphere.

The problem with trying to see the "flex step" in the 3-disc case is discussed in the following section.

§A2 3-DISC FLEX STEP

For the "flex step" in the 3-disc case, we immediately encounter the problem that our 3-dimensional vision does not allow us to see "inside" of a 2-sphere. Look back at Figure 1.3 of Chapter 1 where Freddy has 3-dimensional vision — he can see a 3-disc. But he cannot see any of the points inside of the 3-disc. Recall Freddy with 1-dimensional vision in Figure 1.8 — with a 1-disc completely within his visual line, he cannot see any point *inside* the 1-disc.

The problem is that if we cannot visually see any points inside a 3-disc, then where are the "openings" for moving the "inside points" to the outside of a 3-disc? We need to move the "inside points" to the outside so we can prepare for the gluing step.

The answer is simple — add yet another line ℓ'''' to the lines ℓ, ℓ', and ℓ''' defined in the 2-disc case. The line ℓ'''' *touches* our 3-dimensional visual space at exactly one point.[3]

Created by the line ℓ'''', we now have a forth dimension into which we may move the inside of our 3-disc to "picture" our "flexed 3-disc" (hyper-hemisphere).

In §A5 and §A6 below we shall describe the various slices of a 3-sphere induced by lines (1-dimensional), planes (2-dimensional), and hyperplanes (3-dimensional). The format of the reasoning is that of rational arguments, i.e., *theorems* and *proofs*. A *theorem* is a "true statement" consisting of the first assumptions and claim of a chain of rational thinking. A *proof* is essentially the list of details that describe each link in a chain of rational thinking.

§A3 USING A MIRROR TO GLUE DISCS

In the previous two sections the "flex steps" of gluing various discs were detailed. However, after the "flex step" there was a "gluing step". In this section, we return to the 1-disc case to study an alternative way to view the "gluing step."

---

[3] Recall that the line ℓ''' (with 3 prime symbols "'") introduced in the 2-disc case passed through the "paper" on the table at exactly one point on its way to the ceiling.

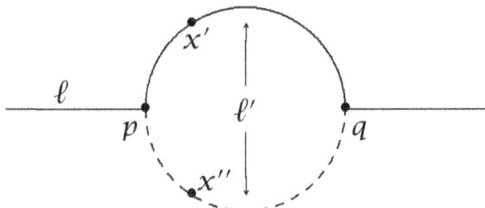

FIG. A1.4 Gluing two semicircles viewed as mirroring one semicircle.

The idea is that of using *a mirror* to produce a *symmetric image* of a semicircle that touches the mirror only at its endpoints. The idea is pictured in Figure A1.4 where the line ℓ is the mirror, the point x'' is the mirror (or symmetric) image of x', and the semicircle (dashed curve) below the line ℓ is the mirror image of the semicircle (solid curve) above the line ℓ. Note that the S^0 boundary points, p and q, are fixed within the mirror ℓ and are thus their own *mirror image.* So the mirror image construction is equivalent to gluing S^0 boundaries.

In the "gluing step" for the 1-disc case the "mirror" is but a single line. In the 2-disc case the "mirror" is a flat plane, i.e., an ordinary typical mirror where the gluing step is that of placing a hemisphere on the surface of the mirror so that *only* the bounding circle touches the mirror.

And for the "gluing step" in the 3-disc case, the "mirror" is a "hyper-mirror" — a *hyperplane of dimension three* living inside a space with four dimensions.

Our hyper-mirror contains the 2-sphere boundary of our 3-disc, but does not contain any of the points inside the 3-disc. In other words, we have moved the inside of our 3-disc to the outside of our mirror using the fourth dimension. The result is a "flexed 3-disc" (*hyper-hemisphere*) whose 2-sphere boundary is contained in the hyper-mirror. This hyper-hemisphere, with its 2-sphere boundary within our mirror, creates a symmetric image, and together the hyper-hemisphere and its "mirror" image equate to gluing two hyper-hemispheres along their 2-sphere boundaries.

§A4 Patterns and the 3-sphere

"Gluing of disc boundaries," Figures 1.5 and 1.6 of Chapter 1, as well as the "slicings of spheres," Figures 1.9 and 1.11 of Chapter 1, provide patterns that extend to the 3-sphere.

For the "gluing pattern," first recall the concept of *disc* introduced in §3:

A 1-*disc* is a line segment that contains its S^0 boundary.

A 2-*disc* is the area inside a circle S^1 that contains its S^1 boundary.

A 3-*disc* is a solid ball that contains its S^2 boundary.

GLUING PATTERN:

> Two 1-discs glued along their S^0 boundaries yield S^1 (Figure 1.5).
> Two 2-discs glued along their S^1 boundaries yield S^2 (Figure 1.6).
> Two 3-discs glued along their S^2 boundaries yield S^3 (no picture).

The "(no picture)" phrase in the previous sentence is a mnemonic for the fact that we cannot see a 3-sphere.

SLICING PATTERN:

> A 1-sphere may be viewed as 0-sphere slices and two points (Figure 1.9).
> A 2-sphere may be viewed as 1-sphere slices and two points (Figure 1.11).
> A 3-sphere may be viewed as 2-sphere slices and two points (no picture).

Again, the "(no picture)" phrase in the previous sentence is a mnemonic for the fact that we cannot see a 3-sphere.

For those readers with an appropriate background, a (formal) rational argument explaining the slicing property for the 3-sphere is provided in the following two sections.

§A5 SLICES OF THE 3-SPHERE

Using a knife that moves only in a fixed plane, a chef can obtain slices of those onions that meet the plane of the knife. Similarly, using a fixed line or fixed plane we may slice those 3-spheres that meet the line or the plane. Our goal here is to determine the shape of the slices using a fixed line, a fixed plane, and the *general equation of a 3-sphere*. In mathematical terms we use the Cartesian product $\mathbb{R}^4$, where $\mathbb{R}$ is the real number line, to denote 4-space — the space of all *tuples* (x_1, x_2, x_3, x_4) that contains all of our 3-spheres. Indeed, within $\mathbb{R}^4$ the *general equation* of a 3-sphere of radius R centered at the point (c_1, c_2, c_3, c_4) is given by

$$(x_1 - c_1)^2 + (x_2 - c_2)^2 + (x_3 - c_3)^2 + (x_4 - c_4)^2 = R^2.$$

The corresponding 3-sphere consists of all points (x_1, x_2, x_3, x_4) that satisfy the equation. In addition, our fixed line ℓ may be viewed as all points of the form $(x_1, 0, 0, 0)$, i.e., all points whose last three components are zero, and our fixed plane π may be viewed as all points of the form $(x_1, x_2, 0, 0)$, i.e., all points whose last two components are zero.

For the theorems (true statements) below, keep in mind that "$\ell \cap S^3$" denotes *the points common to both ℓ and S^3*.

Theorem 1. *Let ℓ be the line of points $(x_1, 0, 0, 0)$ in $\mathbb{R}^4$, and let S^3 be a 3-sphere. Then $\ell \cap S^3$ is a 0-sphere S^0 or a single point, or, ℓ misses S^3.*

PROOF. To find the points in $\ell \cap S^3$ we merely substitute the points of ℓ into the general S^3 equation, thereby obtaining

$$(x_1 - c_1)^2 + (0 - c_2)^2 + (0 - c_3)^2 + (0 - c_4)^2 = R^2$$

or,

$$(x_1 - c_1)^2 = R^2 - c_2^2 - c_3^2 - c_4^2$$

where $R^2 - c_2^2 - c_3^2 - c_4^2$ may be greater than zero, zero, or less than zero. When $(x_1 - c_1)^2 = r^2 = R^2 - c_2^2 - c_3^2 - c_4^2$ is greater than zero, then $x_1 = r + c_1$, and $x_1 = -r + c_1$ are the two solutions, i.e., the equation produces a 0-sphere S^0 of solutions. Next, if $(x_1 - c_1)^2$ is zero, then $x_1 = c_1$ is the only solution, in which case ℓ meets S^3 in exactly one point. And finally, since $(x_1 - c_1)^2$ must be zero or greater than zero, if $R^2 - c_2^2 - c_3^2 - c_4^2$ is less than zero, the equation yields no solutions, in which case ℓ misses S^3. □

Theorem 2. *Let π be the plane of points $(x_1, x_2, 0, 0)$ in $\mathbb{R}^4$, and let S^3 be a 3-sphere. Then $\pi \cap S^3$ is a 1-sphere S^1 or a single point, or, π misses S^3.*

PROOF. To find the points in $\pi \cap S^3$ we merely substitute the points of π into the general S^3 equation, thereby obtaining

$$(x_1 - c_1)^2 + (x_2 - c_2)^2 + (0 - c_3)^2 + (0 - c_4)^2 = R^2$$

or

$$(x_1 - c_1)^2 + (x_2 - c_2)^2 = R^2 - c_3^2 - c_4^2,$$

which is the equation of the circle (1-sphere) in π centered at $(c_1, c_2, 0, 0)$ with radius r when $r^2 = R^2 - c_3^2 - c_4^2$ is positive. But $R^2 - c_3^2 - c_4^2$ may be less than zero, or, zero. The "zero case" yields exactly one solution $(c_1, c_2, 0, 0)$, while the "less than zero" case yields no solutions because the left side of the equation cannot be less than zero, which means π misses S^3. So π misses S^3, or $\pi \cap S^3$ is either a single point or a 1-sphere S^1. □

§A6 HYPERPLANE SLICES OF THE 3-SPHERE

Like our chef who used a knife that moved only in a fixed plane to slice onions, we can use the fixed hyperplane of points $(x_1, x_2, x_3, 0)$ in $\mathbb{R}^4$ to slice 3-spheres. Indeed, to classify the slices we might proceed as we did in the proofs of Theorems 1 and 2: Substitute the points $(x_1, x_2, x_3, 0)$ into the general equation of a 3-sphere and obtain

$$(x_1 - c_1)^2 + (x_2 - c_2)^2 + (x_3 - c_3)^2 = R^2 - c_4^2$$

which is an equation of a 2-sphere centered at $(c_1, c_2, c_3, 0)$ of radius r for $r^2 = R^2 - c_4^2$ whenever $R^2 - c_4^2$ is greater than zero. But if $R^2 - c_4^2$ were negative, then there is no solution; and if $R^2 - c_4^2$ is zero, then there is exactly one solution $(c_1, c_2, c_3, 0)$.

But for those readers who may have an interest in other kinds of proofs, proofs analogous to a chef who uses only one size onion and holds that onion in one particular place while he moves his knife into any position, we shall take another approach. We fix our 3-sphere S^3 as the 3-sphere whose center is located at $(0, 0, 0, 0)$ and whose radius is "1" — our fixed S^3 is the *standard* 3-*sphere* whose points (x_1, x_2, x_3, x_4) satisfy

$$x_1^2 + x_2^2 + x_3^2 + x_4^2 = 1.$$

Then we shall use general formulas for moving 3-dimensional hyperplanes to obtain the desired result. In this case, however, the mathematics in the background moves beyond the high-school level, but may be found in most introductory linear algebra or vector analysis books.

Theorem 3. *Let θ be a 3-dimensional hyperplane in $\mathbb{R}^4$, and let S^3 be the standard 3-sphere. Then $\theta \cap S^3$ is the empty set, a single point, or a 2-sphere.*

PROOF. Suppose the hyperplane $\theta = \theta_0 + q$ in $\mathbb{R}^4$ is the translation $\theta_0 + q$ of a 3-dimensional subspace θ_0 of $\mathbb{R}^4$. There are two cases, according to $q \in \theta_0$ or $q \notin \theta_0$.

In the first case, $\theta = \theta_0$. The Gram-Schmidt Process provides the existence of an orthonormal basis $A = \{e_1, e_2, e_3, e_4\}$ of $\mathbb{R}^4$, such that $\{e_1, e_2, e_3\}$ is an orthonormal basis of θ_0 and e_4 is perpendicular to θ_0. This basis A may differ from the *standard orthonormal* $\mathbb{R}^4$-*basis* $B = \{u_1 = (1, 0, 0, 0), u_2 = (0, 1, 0, 0), u_3 = (0, 0, 1, 0), u_4 = (0, 0, 0, 1)\}$. In addition, the relation between A and B is geometrically the result of a few "rotations" about the origin "$(0, 0, 0, 0)$" common to both A and B. In such a case, when we express a vector p within the A system its length is the same as when we express p within the B system.

Thus any point p in $\theta_0 \cap S^3$, simultaneously in θ_0 and S^3, may be expressed as $p = xe_1 + ye_2 + ze_3 + 0e_4$ and $p = x_1u_1 + x_2u_2 + x_3u_3 + x_4u_4$. Moreover, because p is in S^3 and the length of p is invariant, i.e., is independent of the representation, we have

$$p \cdot p = x^2 + y^2 + z^2 = 1 = x_1^2 + x_2^2 + x_3^4 + x_4^2.$$

So any point p common to both θ_0 and S^3 is a point on a unit 2-sphere inside θ_0 whose center is $(0, 0, 0, 0)$. Conversely, because length is invariant, any point on this unit 2-sphere is clearly a point in S^3.

Turning to the case $q \notin \theta_0$, we know that the point q is not the origin of $\mathbb{R}^4$. In this case we extend our orthonormal basis $\{e_1, e_2, e_3\}$ of θ_0 to an *orthogonal* basis $\{n, e_1, e_2, e_3\}$ of $\mathbb{R}^4$ by defining

$$n = q + a_1 e_1 + a_2 e_2 + a_3 e_3 \quad \text{where each } a_i = -(q \cdot e_i).$$

The key properties of n are the following:

(a) each $n \cdot e_i = q \cdot e_i + a_i(e_i \cdot e_i) = q \cdot e_i + (-q \cdot e_i) = 0$,
(b) $\{n, e_1, e_2, e_3\}$ is an orthogonal basis of $\mathbb{R}^4$, and
(c) because $(n - q) \in \theta_0$, we have a coset equality $\theta_0 + q = \theta_0 + n$
(d) $n \in \theta = \theta_0 + q$

So (a) implies the vector n is perpendicular to the hyperplane θ_0.

Under the translation of θ_0 by n, the e_i points move to $u = e_1 + n$, $v = e_2 + n$, and $w = e_3 + n$, which, from property (c), are points in θ. The four points u, v, w, and n in θ are linearly independent points because the vectors $u - n = e_1$, $v - n = e_2$, and $w - n = e_3$ are linearly independent. So each $p \in \theta$ is a linear combination of these four vectors according to

$$p = xu + yv + zw + tn = xe_1 + ye_2 + ze_3 + n \quad \text{where } t = 1 - x - y - z.$$

It follows that whenever p is in θ, then $p \in \theta \cap S^3$ if and only if $p \cdot p = 1$. Also

$$\begin{aligned} p \cdot p &= (xe_1 + ye_2 + ze_3 + n) \cdot (xe_1 + ye_2 + ze_3 + n) \\ &= x^2 ||e_1||^2 + y^2 ||e_2||^2 + z^2 ||e_3||^2 + ||n||^2 + 2xy(e_1 \cdot e_2) + 2xz(e_1 \cdot e_3) \\ &\quad + 2x(e_1 \cdot n) + 2yz(e_2 \cdot e_3) + 2y(e_2 \cdot n) + 2z(e_3 \cdot n) \\ &= x^2 + y^2 + z^2 + ||n||^2. \end{aligned}$$

So for $p \in \theta$, we have $p \in \theta \cap S^3$ if and only if $x^2 + y^2 + z^2 = 1 - ||n||^2$.
In summary, if $\theta = \theta_0$, then
CASE I: $\theta \cap S^3$ is a unit 2-sphere.
And if $\theta \neq \theta_0$, then
CASE II: $0 < ||n|| < 1$, implies $\theta \cap S^3$ is a 2-sphere of radius less than 1.
CASE III: $||n|| = 1$, implies $\theta \cap S^3$ is a single point n.
CASE IV: $||n|| > 1$, implies $\theta \cap S^3$ is empty (θ misses S^3). □

§A7 Dante's 3-sphere Construct

Our Chapter 2 *Dante's 3-sphere Universe* is based on Mark Peterson's article *Dante and the 3-sphere* published in the American Journal of Physics, Volume 47, in 1979. But on pages 38 and 39 of Donal O'Shea's 2007 book *The Poincaré Conjecture, In Search of the Shape of the Universe* published by Walker and Company, New York, we find the following:

> We have more difficulty imagining the three-sphere as a whole because we don't have an extra dimension to get outside it. In the case of the three-sphere, the hemispheres are not two-dimensional disks with boundaries; they are the two solid balls, and the common boundary is not a circle, but the two-dimensional sphere.
>
> A number of scholars have argued convincingly that the universe imagined in *The Divine Comedy* by the great Italian poet and writer Dante Alighieri (1261-1321) is a three-sphere (although he did not, of course, call it that).[30]

The superscript "30" in the quote above refers to "Note 30" on page 206 of O'Shea's book:[4]

> See M. Peterson, "Dante and the 3-sphere," ..., and R. Osserman, *poetry of the Universe* (Garden City, NY: Doubleday, 1995). Apparently, this was noticed many years ago by the mathematician Andreas Speiser in his book *Klassische Stücke der Mathematik* (Zürich: Verlag Orell Füselli, 1925). The latter is referenced in the article by J.J. Callahan, "The Curvature of Space in a Finite Universe," *Scientific American* 235 (August 1976): 90-100. I owe these references to M. Peterson.

---

[4] "Note 30" tells us that in addition to Mark Peterson, others noticed that the Dante construct is a 3-sphere. This author's awareness of these historical references is a direct result of Eugene Miller's interest in this book. So it is important to document Eugene's contribution.

CHAPTER 13

Appendix 2: Supplement for Chapters 3 and 4

This Appendix extends, in the sense of providing details and technical information, Chapters 3 and 4.

Part I: Supplement for Chapter 3

§A8 NON-EUCLIDEAN GEOMETRY

A good reference for non-Euclidean geometry is John Stillwell's book *Yearning for the Impossible* published by A. K. Peters, Ltd. Wellesley, Massachusetts, Copyright © 2006 by A K Peters, Ltd.

We begin with the word *geometry*, which breaks into two parts *geo* meaning "earth" and *metry* referring to "measurement." When humans first began dividing land, they needed measurements to mark off areas on the surface of the earth. And anything that is used over and over is usually simplified, the ultimate simplification is generally viewed as *axiomatics* — a list of rules that are to be logically, and therefore consistently, applied. Euclid (circa 300BC) taught geometry as a logical subject in Alexandria, where he gathered together what was then known of geometry into thirteen books which he titled the *Elements*.

The phrase *Euclidean geometry* refers to the mathematics contained in the Elements, and is the geometry still taught in secondary schools. Euclidean geometry is sometimes referred to as *plane geometry*.

Historically, the human view of the earth was that of a flat (no curvature) plane. And so the axioms of Euclidean geometry captured the measurements relative to an infinite plane that had no curvature.

In such a flat-plane geometry there exist *parallel lines*, lines that never intersect. But when we consider the surface of a 2-sphere S^2 the *lines are great circles*, circles that contain *antipodal points* (the north and south poles provide an example of a pair of antipodal points). And every pair of great circles on a 2-sphere intersect. So "lines (great circles) on S^2" do not satisfy

---

Electronic supplementary material The online version of this chapter (doi: 10.1007/978-3-319-06254-9_13) contains supplementary material, which is available to authorized users. Videos can also be accessed at http://www.springerimages.com/videos/978-3-319-06253-2

the axioms of Euclidean geometry, and a study of such a curved surface is a study of a *non-Euclidean geometry*.

Roughly speaking, we may compare Euclidean and non-Euclidean geometries with the following thought experiment: Place a round stake in the ground with a loose-loop of string around the stake and a fixed length "r" of string attached to the loose-loop. Then as you walk around the stake holding the fixed-length string taut you can mark the ground and thereby create a circle. If the ground surrounding your stake is perfectly flat (no curvature), then Euclidean geometry tells you that the *perimeter* (length) of the circle is "$2 \times \pi \times r$" where "r" is the *radius* and "π" (pi) is an anciently (circa 1650BC) famous number.[1]

But if you try this experiment on a *curved surface*, for example a sphere or a saddle surface as illustrated in §4, and if your string is long enough and both the sphere and saddle surfaces are large enough, then the perimeter of the circle will not match the predicted perimeter "$2 \times \pi \times r$" calculated within Euclidean geometry. So the geometry of these surfaces is non-Euclidean.[2]

§A9 CONTEXT OF EINSTEIN'S QUOTES

Except for the first sentence in Einstein's quote in §14, namely, *From the latest results of the theory of relativity it is probable that our three-dimensional space is also approximately spherical ...*, his quotes were presented without motivation concerning his particular reason why he would be interested in a 3-sphere, i.e., his quotes were presented without context.

The context for his quotes in Chapter 3 is the distinction between a non-Euclidean 3-sphere and ordinary Euclidean 3-dimensional space. This Euclidean space is *infinite* while the 3-sphere is *bounded*. The context is the distinction between "infinite" and "non-infinite" spaces. Comparing the 3-sphere to 3-dimensional Euclidean space is like comparing a circle, which has a *finite* length (perimeter), to a line, which has *infinite* length. Or comparing a 2-sphere, which has finite area, to a plane, which has infinite area.

But the best way to provide context for Einstein's quotes is to again quote Einstein. So in Einstein's *Ideas and Opinions*, we visit page 233:

> ... the question whether the universe is spatially finite or not seems to me an entirely meaningful question in the sense of practical geometry.

---

[1] The number π is approximately 3.1459 $\cdots$, where the "$\cdots$" indicate the digits continue indefinitely. The first 8 billion digits of π have been calculated — see David Blatner's 1997 book *The Joy of pi* published by Walker and Company, NY — [5] in Bibliography. For an example of applying the *perimeter formula*, note that the perimeter of a circle with a 10-foot radius is $2 \times (3.1459 \cdots) \times 10 = 62.981 \cdots$ feet.

[2] The *perimeter* of a circle of radius r on flat ground will be *greater than* the perimeter of a circle of radius r measured on the sphere and *less than* the perimeter of a circle of radius r measured on the saddle surface. For a concise overview of surfaces, see Chapter 8 *Riemann's Legacy* of O'Shea's book, *The Poincaré Conjecture* — reference [26] in Bibliography.

> I do not even consider it impossible that this question will be answered before long by astronomy. Let us call to mind what the general theory of relativity teaches in this respect. It offers two possibilities:
>
> 1. The universe is spatially infinite. This is possible only if in the universe the average spatial density of matter, concentrated in the stars, vanishes, i.e., if the ratio of the total mass of the stars to the volume of the space through which they are scattered indefinitely approaches zero as greater and greater volumes are considered.
>
> 2. The universe is spatially finite. This must be so, if there exists an average density of the ponderable matter in the universe which is different from zero. The smaller that average density, the greater is the volume of the universe.
>
> I must not fail to mention that a theoretical argument can be adduced in favor of the hypothesis of a finite universe. The general theory of relativity teaches that the inertia of a given body is greater as there are more ponderable masses in proximity to it; thus it seems very natural to reduce the total inertia of a body to interaction between it and the other bodies in the universe, as indeed, ever since Newton's time, gravity has been completely reduced to interaction between bodies. From the equations of the general theory of relativity it can be deduced that this total reduction of inertia to interaction between masses — as demanded by E. Mach, for example — is possible only if the universe is spatially finite.

We couple the Einstein quote concerning a finite universe with a quote from Lincoln Barnett's book *The Universe and Dr. Einstein* pages 97 and 98: (Additional comments appear inside square brackets "[·]".)

> Like most of the concepts of modern science, Einstein's finite, spherical universe cannot be visualized — any more than a photon or an electron can be visualized. But as in the case of the photon and the electron its properties can be described mathematically. By taking the best available values of modern astronomy and applying them to Einstein's field equations, it is possible to compute the *size* of the universe. In order to determine its radius, however, it is first necessary to ascertain its curvature. Since, as Einstein showed, the geometry or curvature of space is determined by its material content, the cosmological problem can be solved only by obtaining a figure for the average density of matter in the universe.

Fortunately this figure is available, for astronomer Edwin Hubble of the Mt. Wilson Observatory conscientiously studied sample areas of the heavens over a period of years and painstakingly computed the average amount of matter contained in them. The conclusion he reached was that in the universe as a whole there is

$$.0000000000000000000000000000001$$

gram[s] of matter per cubic centimeter of space. Applied to Einstein's field equations this figure yields a positive value for the curvature of the universe, which in turn reveals that the radius of the universe is 35 billion light years or

$$210,000,000,000,000,000,000,000 \text{ miles.}$$

Einstein's universe, while not infinite, is nevertheless sufficiently enormous to encompass billions of galaxies, each containing hundreds of millions of flaming stars and incalculable quantities of rarefied gas, cold systems of iron and stone and cosmic dust. A sunbeam, setting out through space at the rate of 186,000 miles a second would, in this universe, describe a great cosmic circle and return to its source after a little more than 200 billion terrestrial years.

Part II: Supplement for Chapter 4

§A10 DOPPLER SHIFT

Within Figure A2.1 we see two corks floating in a pool of water.

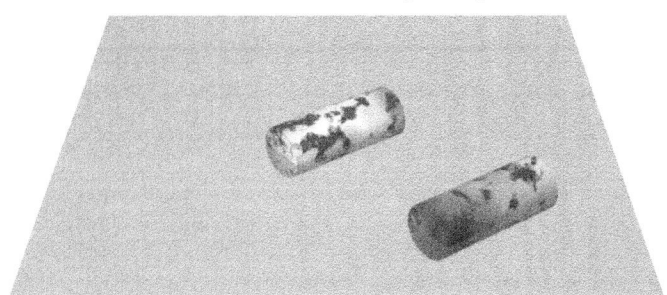

FIG. A2.1 Two corks floating in a pool of water.

Suppose that we cannot see the water, just the two corks. If we push the first cork toward the second, then the second cork moves. Since we do not see the water we would surmise that there is *some kind of interaction* between them. In reality we simply use one cork to push the water which then pushes the other cork.

Next, insert two small pieces of lead into the corks — one piece into each cork so the corks stand vertically within the (invisible) water.

FIG. A2.2 Cork #1 oscillates, producing waves that rhythmically impact cork #2.

Then we *rhythmically* push the first cork up and down, while doing nothing to the second cork. The rhythmic motion creates *waves* — waves that we cannot see but nevertheless move in a circular pattern away from the first cork (Figure A2.2).

The water waves consist of peaks and valleys, and the distance between two consecutive peaks is called the *wave length*. There are also other measures associated with waves. As the waves move past the second cork, the second cork experiences an up and down oscillation. The second cork encounters an *oscillatory influence* that has a certain *frequency*, say for example 15 "peaks" in 20 seconds.

Now let us see what happens to the waves as we *rhythmically push the first cork up and down while simultaneously moving it toward the second cork*. The question is, What happens to the frequency?

A snapshot of the result is illustrated in Figure A2.3 where we see an oscillating cork moving in the direction of the arrow. Within the graphic we also see waves issuing from the moving cork — one square slice of the wave issuing in the *forward direction* of the arrow and another square slice issuing in the *backward direction*.

The key observation is the *difference* between the frequencies of the waves within the two square slices — the frequency of the waves (right-side-square) moving in direction of the arrow is greater than the frequency of the waves (left-side-square) moving in opposite direction. This phenomenon is the *Doppler effect*.

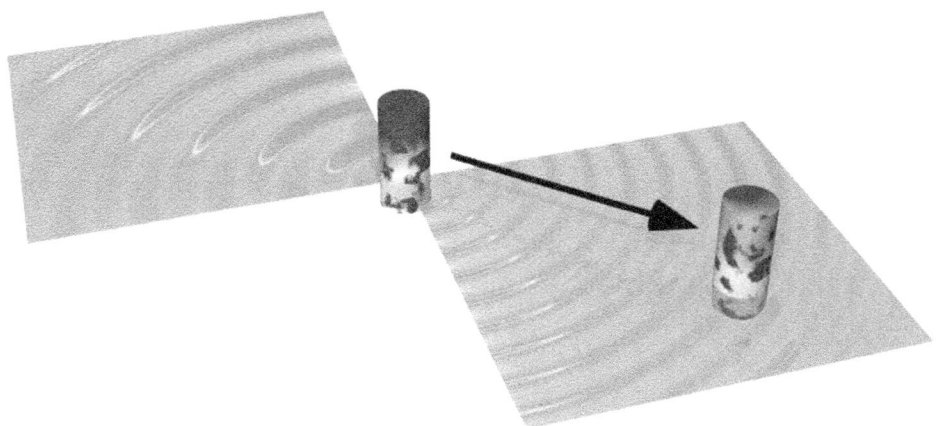

FIG. A2.3 The oscillating cork in the center moves toward the other cork.

So we can see waves moving across the *field of water*. But we cannot easily see a *field of air*, and thus we cannot see waves of rhythmically oscillating air pressures. We do, nevertheless, experience such waves as sounds.

Sound waves are composed of pulses of relatively high air pressure. For example, for a single pluck on a guitar string the motion of the string *jiggles the air*, creating oscillatory changes in air pressure, similar to our jiggling cork that created oscillatory changes in water pressure. These oscillatory guitar string "blasts" travel as "waves" through the air and impact our ear drums similar to how our water waves impacted our "stationary cork". The oscillatory influence on our ear drums is then transformed, via the brain, into the "sound" of a musical note. The rule is *the higher the frequency the higher the pitch.*

Within a field of air we may illustrate sound waves produced by a moving police siren (Figure A2.4).

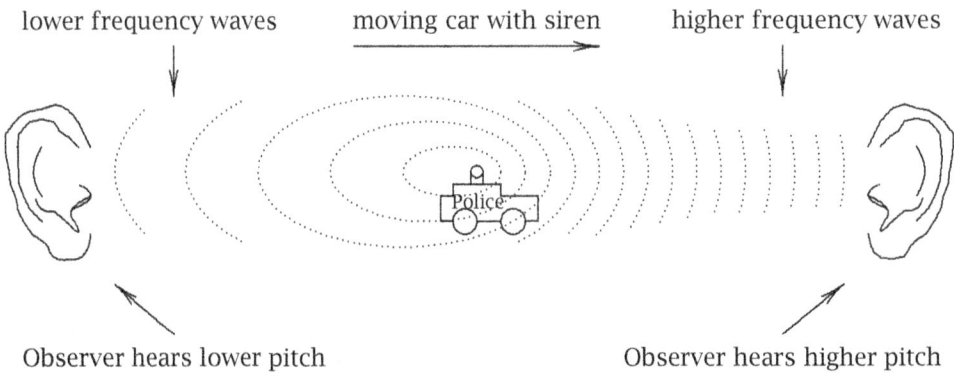

FIG. A2.4 Frequency change adjusts the pitch of what we hear.

§A11 Red shift[3]

Both the *field of water* and the *field of air* serve to propagate waves, waves induced by "jiggles" within the fields. In the *electrical case*, the analogue of the "field of water" is called the *electromagnetic field*. The electromagnetic field carries waves when we "jiggle" a *charge* within the field. An example of a "jiggled charge" is our household alternating current, which alternates its direction at a frequency of 120 oscillations per second.

At higher frequencies electromagnetic waves are experienced as *light* — within limitations of certain frequencies, we use our eyes and brain to experience electromagnetic waves as various colors of light. For example, if the frequencies are at least *five-hundred trillion oscillations per second*, that is,

$$500,000,000,000,000 \quad \text{oscillations per second,}$$

but less than 10 times this number — less than *five quadrillion oscillations per second*, that is, less than

$$5,000,000,000,000,000 \quad \text{oscillations per second,}$$

then humans experience these electromagnetic waves as light — *red light* near the lower end of this spectrum of frequencies and *blue light* near the upper end. But if the frequencies go too high, e.g., above the frequency of X-rays, then human vision cannot detect the "light."

When humans distinguish *red light* from *blue light* they are distinguishing lower frequency (red light) electromagnetic waves from higher frequency (blue light). And analogous to the two observers in Figure A2.4, the Doppler effect must be considered. In the case of the Doppler effect of frequencies of electromagnetic waves, however, we are interested in *galaxies* moving away from earth.[4]

An observer on or near the earth who observes a galaxy moving away from earth will notice over time that the frequency of the electromagnetic waves coming from the galaxy is declining. This decline in frequency measured over time is called a *red shift* — the frequency *shifts* to lower frequencies, like when we *shift* our observations from blue light to red light.

---

[3] The approach used here —corks and water — to motivate the idea of *waves* and the *electromagnetic field* is a variation of a few paragraphs on page 31 of Richard Feynman's book *Six Easy Pieces*, copyright 1995 by the California Institute of Technology. In passing, we note that another standard approach to the idea of *red shift* is based on *wave length*, distance between consecutive peaks. The connection is that longer wave lengths amount to lower frequencies and shorter wave lengths to higher frequencies.

[4] A *galaxy* is a grouping of billions of *stars* (our sun is an example of a star but stars in general may be much larger or much smaller than our sun). And of course galaxies may also contain planets and other matter. A galaxy shaped like a gigantic deformed ball may span a hundred thousand light years or more. (A *light year* is the distance that light travels in one year, and through a vacuum, light travels 186,000 miles per second.)

The idea of a *red shift of a galaxy* is simply a way to convey that observations involving the frequency of electromagnetic waves from the galaxy have been recorded and the change in frequency indicates that the galaxy is moving away from the observer.

Relative to our universe, we can briefly elucidate some of the history of the redshift by quoting Daigneault and Sangalli:[5]

> What is the shape of space? Is the universe finite or infinite? Did the world have a beginning, or has it always existed? These fundamental questions have intrigued and baffled humans since the most ancient times. But it was only in the twentieth century, with the development of powerful tools for probing the immensity of the skies, that it became possible to explore the cosmos beyond our galaxy's neighborhood. Concurrently, advances in physics and mathematics provided the conceptual framework — the language, so to speak — in which to formulate comprehensive theories whose validity could be objectively put to the test. Scientific answers to the above questions finally appeared to be at hand. ...
>
> In 1929 Edwin Hubble interpreted the galactic redshift phenomenon detected by Vespo Slipher starting in 1912—the change in the observed frequency of light waves— as a Doppler effect, that is, as caused by the motion of a luminous source away from the observer. The Doppler interpretation became known as the Expanding Universe theory, whose most developed form is the cosmology of Friedmann and Lemaître and according to which galaxies are moving away from each other. As told in 1997 by David Dewhirst and Michael Hoskin in *The Cambridge Illustrated History of Astronomy*, "There is no doubt that the nearly simultaneous detection of the redshifts and the derivation of solutions of Einstein's equations that suggested that the universe would be expected to expand greatly encouraged this interpretation."
>
> The Expanding Universe theory later begot the Big Bang theory, which maintains, in addition to universal expansion, that time and the universe had a beginning, the "Big Bang".

---

[5]See the article *Einstein's Static Universe: An Idea Whose Time Has Come Back* by Aubert Daigneault and Arturo Sangalli published in the January 2001 issue of the Notices of the American Mathematical Society.

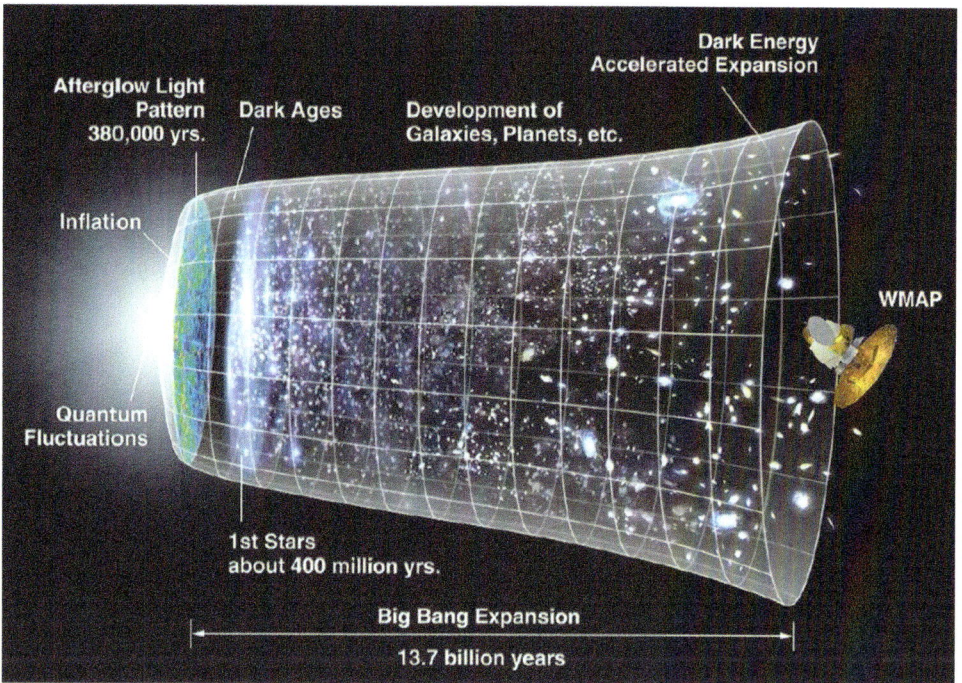

FIG. A2.5 The Big Bang and the Timeline of the Universe. The web site http://map.gsfc.nasa.gov/media/060915/060915_CMB_Timeline150.jpg is the original source of the graphic in this figure. The white-lettered "WMAP" located on the right side of the graphic is an acronym for Wilkinson Microwave Anisotropy Probe. Background for the NASA five-year project and related information may be found at http://map.gsfc.nasa.gov/news/index.html

§A12 BIG BANG

The Big Bang theory is a key part of our Daigneault and Sangalli quote.[6]

The *Big Bang* theory is a theory of how the universe came into being — literally with a big bang. From the most simple of viewpoints, some 13.7 billion years ago there was an extremely hot, extremely small, and extremely dense "dot" — the dot exploded. The reasoning behind the Big Bang is like running a film backwards — if the cosmological red shift is truly an indication of expansion of the universe, then, by running the film of the 13.7 billion years of expansion backwards, we see that everything must have been close at the "beginning."

---

[6]See footnote 5 above.

§A13 Cosmic Background Radiation

Continuing with more spade work before we return to quoting the Daigneault and Sangalli article, we consider the *Cosmic Background Radiation* (CBR).

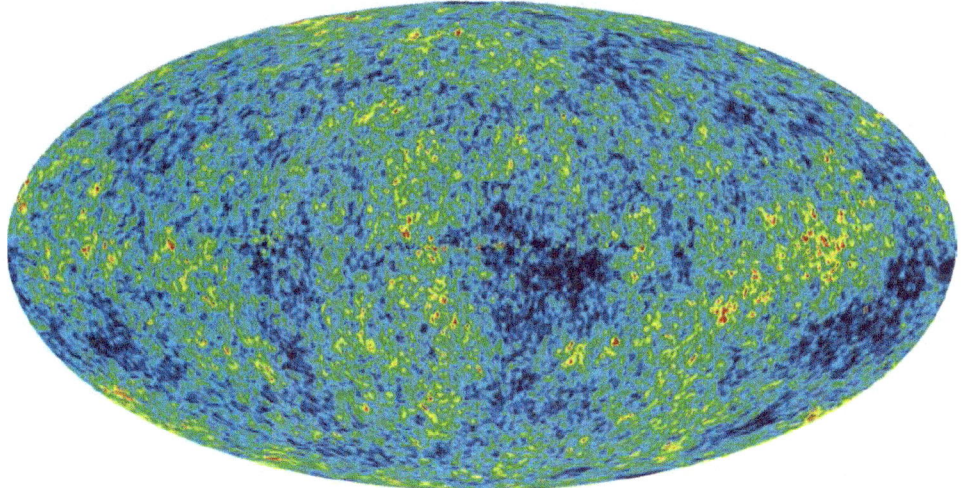

FIG. A2.6 Cosmic Background Radiation —NASA/WMAP Science Team.

In Figure A2.6 we see the full sky 5-year WMAP data of the cosmic microwave temperature fluctuations. The data also appears, from a skewed view, in the Big Bang illustration Figure A2.5 as the *Afterglow Light Pattern 380,000 years*.

The full sky map usually appears in color: Analogous to a weather map, the "colors" represent tiny temperature fluctuations with the lighter regions indicating warmer temperatures and the darker regions colder temperatures. The fluctuations are measured to be about $.00036°F$. The average temperature is $-455°F$.

With the pictures A2.5 and A2.6 of the Big Bang and CBR, let us now return to our quote of the Daigneault and Sangalli article that concerns Einstein's Universe (EU), the topic of Chapter 4.

§A14 Concerns about Einstein's Universe

We return to quoting the Daigneault and Sangalli article:

> The Expanding Universe theory later begot the Big Bang theory, which maintains, in addition to universal expansion, that time and the universe had a beginning, the "Big Bang".
>
> Hubble also stated his famous law: The galaxies recede from each other with a velocity proportional to the distance separating them. But he did

not rule out that, by virtue of some unknown mechanism, the redshift might result from space being curved.

When speaking of Hubble's famous law — last paragraph of the quote above — let us focus on the sentence, *But he did not rule out that, by virtue of some unknown mechanism, the redshift might result from space being curved.* Said differently, the redshift data did not rule out the possibility that the curvature of EU could somehow be causing the redshift.

Continuing with the quote, we nevertheless find the following:

> ... Once Hubble's expansion seemed to have been established as an empirical fact, Einstein was forced to abandon his static universe model.
>
> The observation in 1964 of the so-called cosmic background radiation (CBR) is often considered to be the smoking gun supporting the assumption that there was a Big Bang. But in fact the CBR might have other causes. Erwin Finlay-Freundlich and Max Born had already predicted the existence of such radiation in 1953, eleven years before its detection, on the basis of a stationary universe. The interpretation of the CBR as a faint echo of the birth of the universe gave greater credence to the Big Bang theory, which became widely accepted as the standard astronomical gospel.

The first part of the last paragraph in the quote above speaks to the cosmic background radiation (CBR) as the ... *smoking gun supporting the assumption that there was a Big Bang.* And then this "Big Bang supporting view" is countered with a statement that the Big Bang could be supported *on the basis of a stationary universe.*

The competing "point counter-point" views of the "Big Bang" verses a "stationary universe" are judged by stating that the Big Bang theory has become widely accepted as the standard astronomical gospel. But as we shall see in the next section, the argument is evidently not over.

§A15 CHRONOMETRIC COSMOLOGY[7]

We continue with our quote which now questions the "Big Bang cosmology."

> By the early 1970s the rapidly increasing mass of statistical data on quasars and galaxies provided a substantial basis for questioning the Big Bang cosmology. In 1972 Irving Ezra Segal, a professor of mathematics at the Massachusetts Institute of Technology, picked up where Einstein had left

---

[7] *Chronometric* contains "chrono" meaning "time", and "metric" meaning "measure".

off and proposed a variant of special relativity[8]: chronometric cosmology (CC), so called because it is based on the analysis of time. His 1976 book *Mathematical Cosmology and Extragalactic Astronomy*[9] contains a detailed presentation of the theory.

And then the next quote tells us why the title of the article *Einstein's Static Universe: an Idea Whose Time Has Come Back?* is what it is.[10]

> According to CC, Einstein's model is the correct one to understand the universe as a whole (i.e., global space-time), except that there are two kinds of time: a cosmic or Einstein's time t, and a local or Minkowski's time x_0, which is (perhaps!) the time measured by existing techniques.

So it is now 1976, and Segal's *chronometric cosmology* theory (CC) puts Einstein's Cartesian product $R \times S^3$ back into the possible pictures of our universe. Let us continue with the short description of CC:

> ... In Segal's words, "the key point is that time and its conjugate variable, energy, are fundamentally different in the EU from the conventional time and energy in the local flat Minkowski space M that approximates the EU at the point of observation." Simply put, Einstein's cosmic time t is the "real" one, whereas Minkowski's time is only an approximation of t.

§A16 FINAL PARAGRAPH OF QUOTE

The remainder of the article that tracks Einstein's Universe $R \times S^3$ concerns the chronometric cosmology theory (CC) and how it addresses other cosmological theories. To close the quotes, so to speak, we include the final paragraph of the article:

> Hence we may conclude on a rather speculative but perhaps conciliatory note. Assuming CC, if the concept of the totality of the matter dispersed in space S^3 as a "gas of galaxies", i.e., a gas the molecules of which are the galaxies, is valid, as is generally taught, then this gas is almost always in equilibrium. However, if Poincaré's theorem is applicable to it, on rare scattered occasions, yet infinitely many times in past cosmic eternity, it has taken very implausible configurations giving rise to fairly big "bangs", and the same would be in stock for the future.

---

[8]The proposed "variant of special relativity" appears in Segal's article *A variant of special relativity and long-distance astronomy* Published in the Proc. Nat. Acad. Sci., Vol. 71 in 1974, pages 765 to 768.

[9]Published by Academic Press, New York, 1976.

[10]Within the quote, note the use of *Minkowski's time* — in the context of Minkowski's space which is the Cartesian product $R \times R^3$, Minkowski's time is the first factor R which is the Real timeline. The second factor R^3 may be viewed as our ordinary 3-dimensional human-vision space.

§A17 PARALLEL UNIVERSES AND GOD THE OBSERVER

There are theoretical problems with the idea of a Big Bang — the problem is that the large scale theory (Einstein's Relativity theory of gravity) and the small-scale theory (quantum theory) need to be considered as parts of a single theory that includes both as consistent subparts.

One such theory is the *Parallel Universes* theory. In his *The First Observer of the Big Bang* Chapter in his *Parallel Universes* book, Fred Alan Wolf discusses what is needed to start the Big Bang:[11]

> The Bible informs us that in the start-up condition there was, first of all, God. And then there was this division of waters. And at some point God let there be light and there was light. Farther on we read that God began to name His manifestations. Each act of naming a thing, thus, in a sense, created that thing as a reality. God's role was, then, that of both the creator and the observer of all things. It would appear, therefore, that perhaps creation and observation (wherein a thing is seen as a thing and called a name) are equally important at the beginning of time.
>
> The idea turns out to be important in quantum physics, something I have called in my earlier books *the observer effect*; and it is also important in attempting to describe the quantum physics of the early universe and the role that parallel universes played when things began.
>
> The *observer effect* is the sudden change in the probability of observing some property of matter, such as its location in space, when that observation is actually carried out. If no observation actually takes place, the thing remains without location in space. It is said to be in a probability or quantum wave pattern that I call a *qwiff*.
>
> The well-known wave-particle duality of quantum physics is an example illustrating the observer effect. If the observer chooses to observe the wave nature of matter, he or she gives up or has no hope of observing the particle like nature of matter, and vice versa. Thus, what is observable depends on the choice made by the observer. In this sense the observer has much input into the physical properties of the matter observed.

And later in his book, in his Chapter *Who Saw What When?*, Wolf applies the idea of an observer determining reality:

> Thus Hawking brings us back to the question of the first observer. Without this observer, it is not clear how our universe began. In fact, it is increasingly clear that such a question cannot be decided. If our universe

---

[11]Wolf's book *Parallel Universes* was published in 1988 by Simon & Schuster Paperbacks.

is really in a state of energy — ground or excited — then it must, following the atomic analogy, exist in an infinite number of parallel *position* universes.

Thus God, being the first observer, and saying, *Let there be light*, was probably misquoted. God more than likely said, *Let there be energy*, and with those words our universe, and all of the other infinite *position* universes, appeared simultaneously, filling *all-space* in *all-time*.

§A18 UNIVERSAL UNIVERSE

As one might suspect, the existence of parallel universes creates many possibilities for speculation. Indeed, in his 2003 book *The Universe and Multiple Reality* M. R. Franks speaks of a *superuniverse*.[12]

Franks' superuniverse is a *universe* itself, a universe that contains all of the "parallel universes". He defines *energy states* (static universes) as either *contiguous* — two such static universes differ by just one quantum transition — or *noncontiguous* — two such static universes differ by more than one quantum transition.

For an analogy, he uses a two-dimensional sheet of paper (the superuniverse) filled with hexagons (each hexagon a static universe) drawn such that each hexagon is contiguous to six other hexagons, and then points out that the superuniverse structure of *contiguous universes* involves two dimensions. Franks calls attention to the fact that one may speculate that a superuniverse might have 10^{98} dimensions, which he says is *as many dimensions as there are subatomic particles in our three dimensional universe.*

Within the context of *dimension theory*, the mathematical area that was initiated circa 1900 AD when a rigorous definition of *dimension* was formulated, a superuniverse may be viewed as a type of *universal space*.

One such universal space rests on the construction of a space that we shall call J. To think about the notation, consider that the 2-web, 3-web, and 4-web constructs are special cases of the general one-dimensional structure J. For example, the 2-web involves a basic triangle with three corners, and so we use $J = J_3$ to speak about the 2-web. Similarly, we use $J = J_4$ to consider the 3-web with its four corners, and we use $J = J_5$ when we are talking about the 4-web with its five corners. The list continues indefinitely.[13]

---

[12] Franks' book is published by iUniverse, Inc., 2021 Pine Lake Road, Suite 100, Lincoln, NE, 68512.

[13] Ivan Ivanšić and Uroš Milutinović showed that J_3 is sufficient for all of these cases. See Theorem 14 in the *The Quest for Universal Spaces in Dimension Theory* by Stephen Lipscomb. Also see his article in the AMS Notices, Vol. 56, Number 11, December 2009.

Now if a superuniverse exists as a *metric space* (see §A20 Appendix 3) of dimension d, then the superuniverse is a part of the *Cartesian product* space J^{d+1} with $d + 1$ factors J (for Cartesian product recall §19).

So in a sense J^{d+1} is a *Universal universe* — every conceived d-dimensional superuniverse is inside of J^{d+1}. For example, if a superuniverse does indeed have dimension 98, then it may be viewed as a part of J^{99}.

God's Image? Created with 21-Century mathematics as *art from the fourth dimension*.

Michelangelo's *Creation of Adam* (see page 85).

God's Image? left-section of front view

God's Image? right-section of front view

Images (partial and complete) of Great 2-spheres (See Chapter 11)

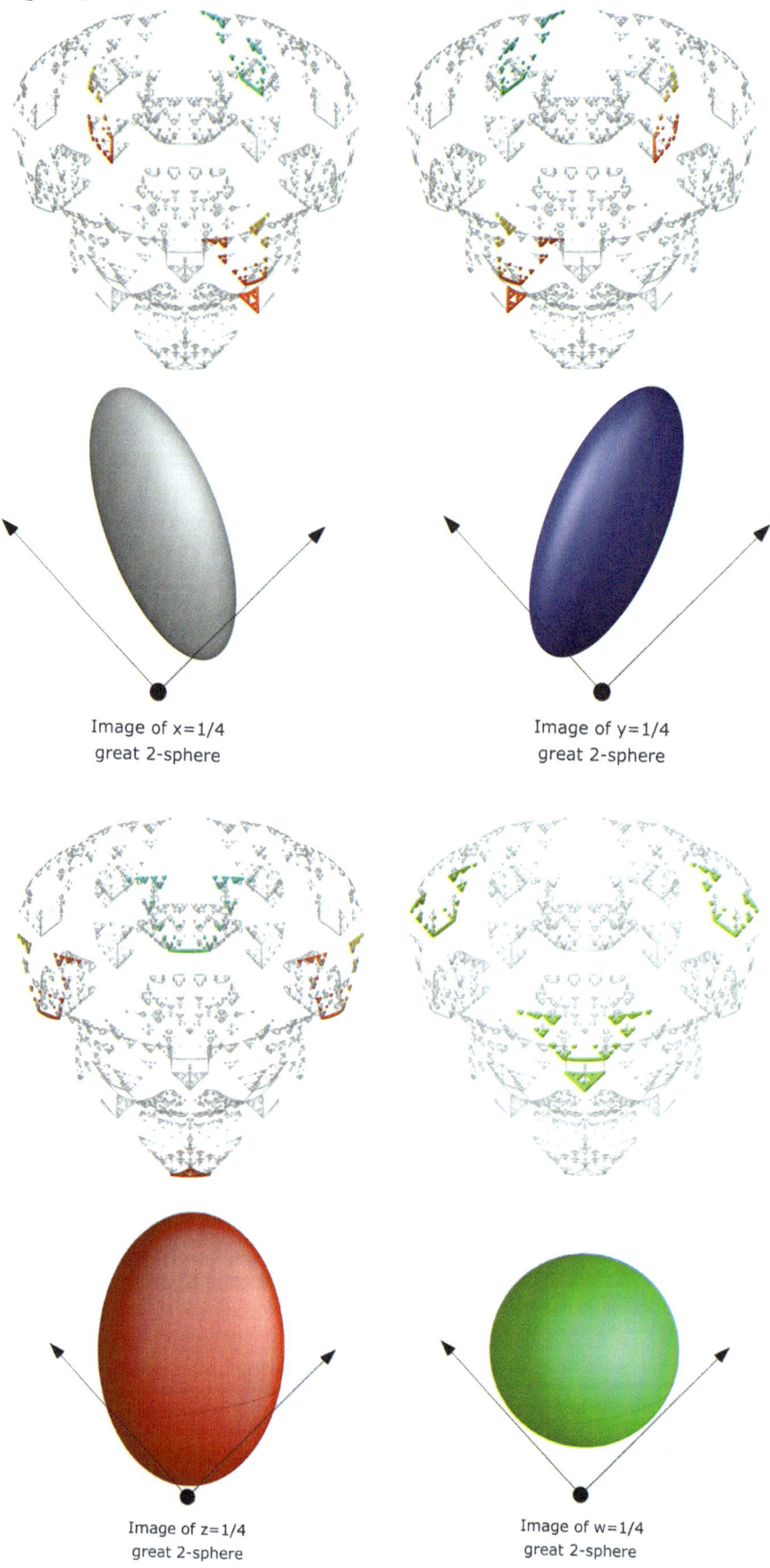

Images of Joins of Great 2-spheres

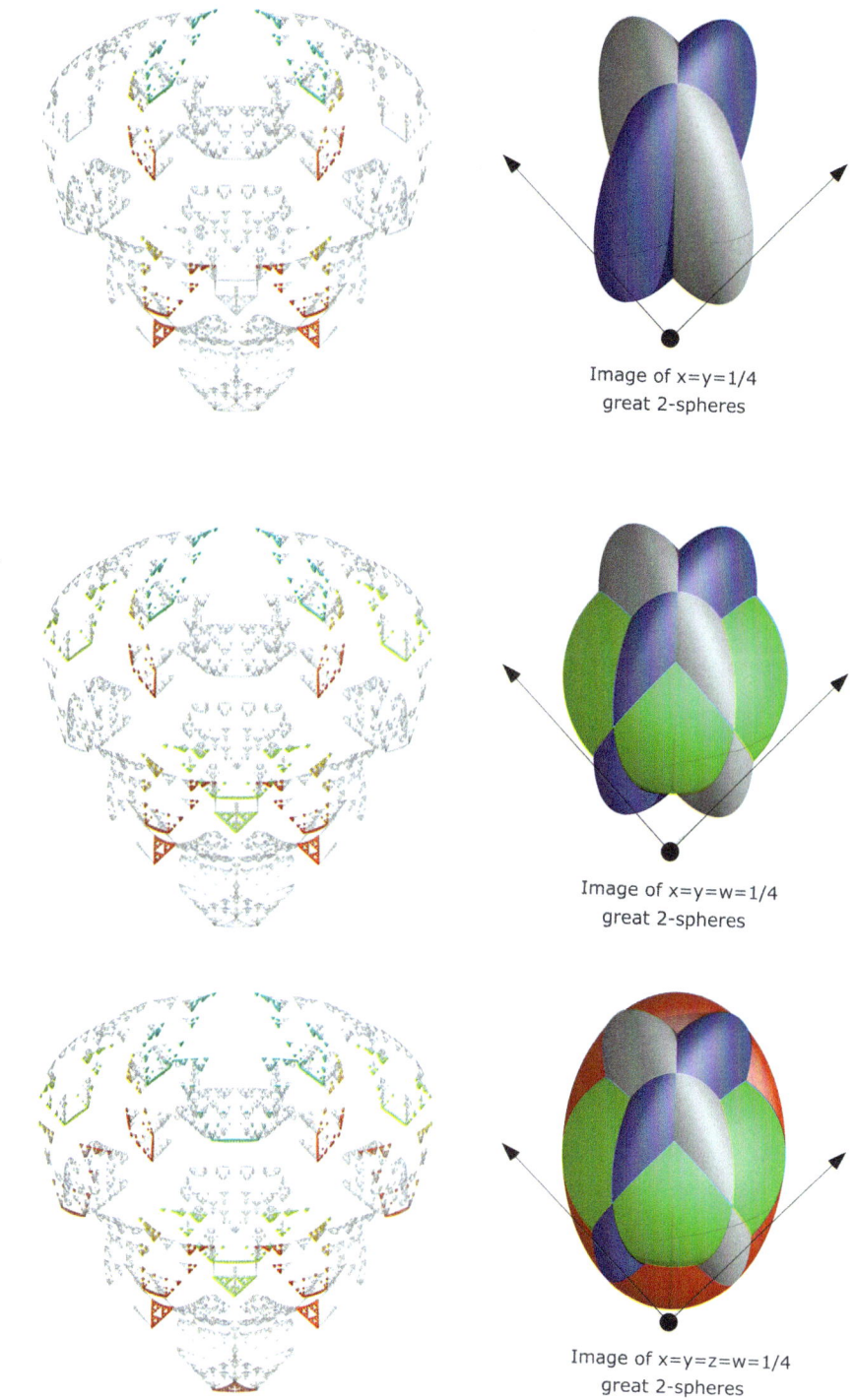

Image of x=y=1/4
great 2-spheres

Image of x=y=w=1/4
great 2-spheres

Image of x=y=z=w=1/4
great 2-spheres

Each point in any **God's Image?** graphic is the image of exactly one point (a single pre-image) on the 4-web grid in the fourth-dimension (see §58). In contrast, the right-side column illustrates *ellipsoidal-join images* of great 2-sphere joins. These ellipsoidal-join images may contain points that have *more than one pre-image* in our hypersphere (§60 and §61).

God's Image? left-section of bottom view

God's Image? right-section of bottom view

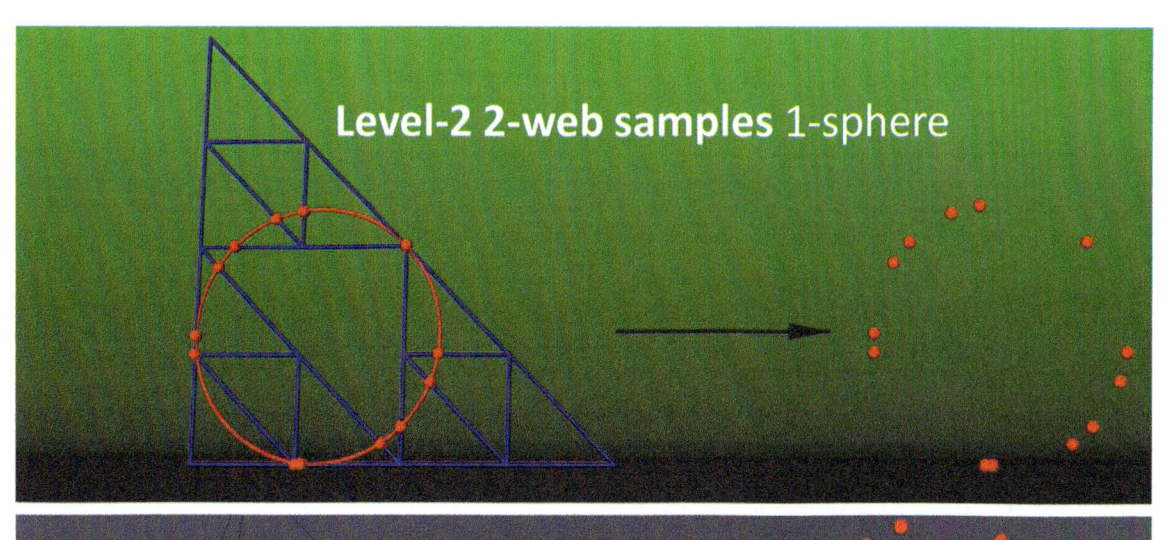

Level-2 2-web samples 1-sphere

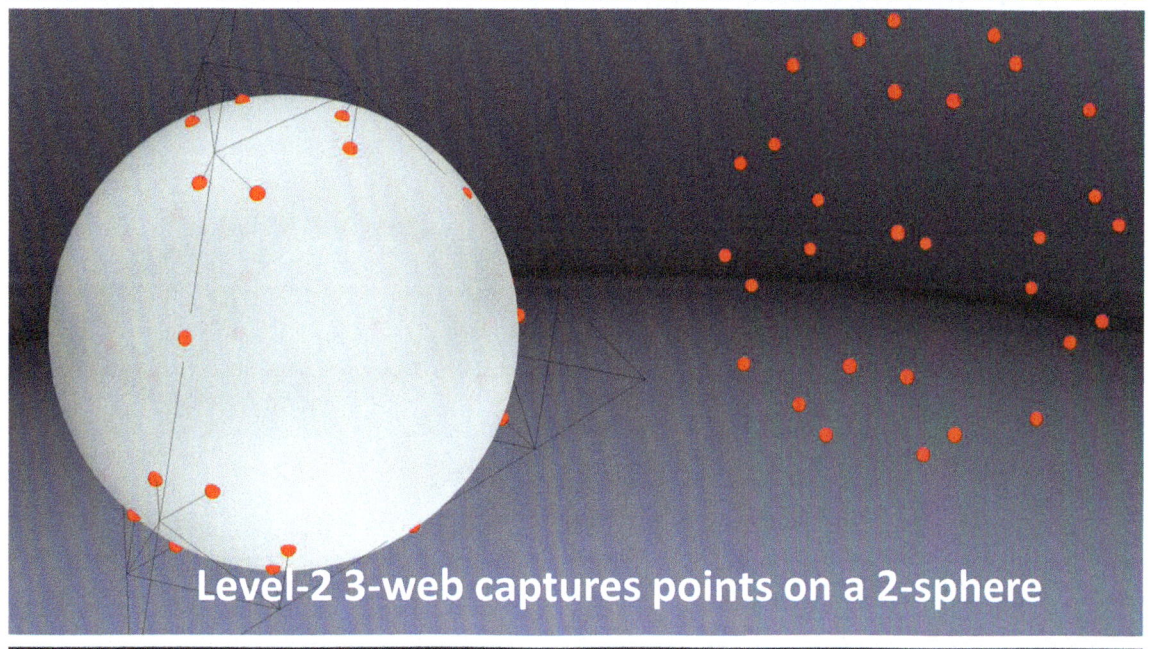

Level-2 3-web captures points on a 2-sphere

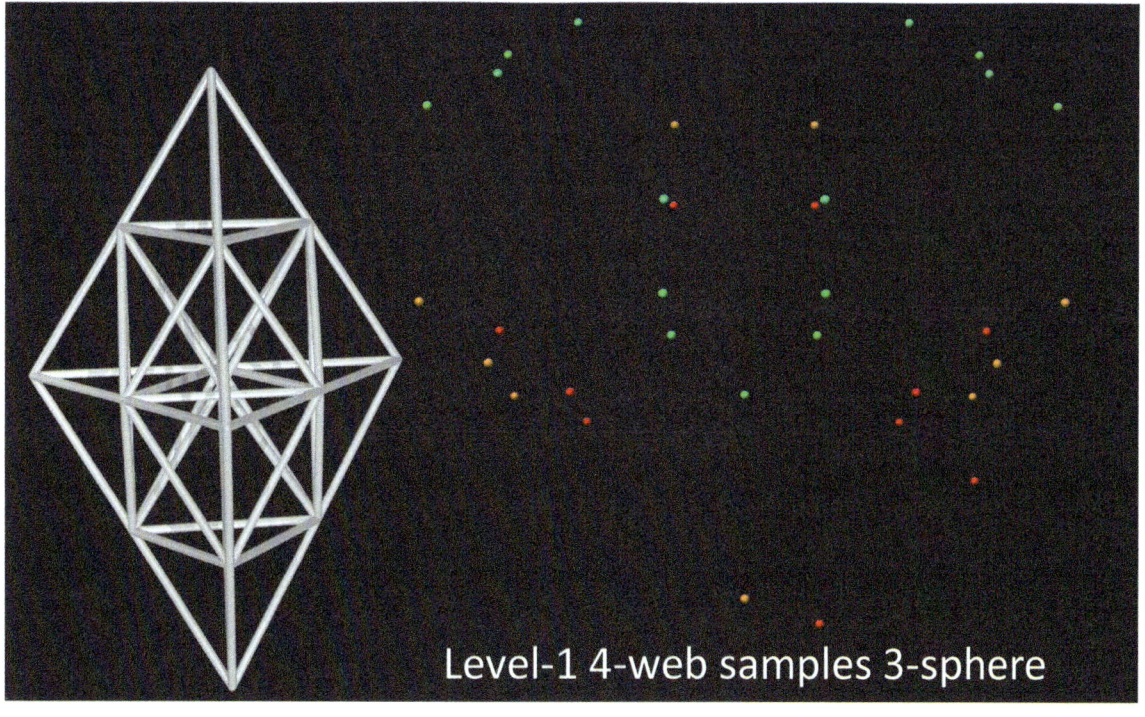

Level-1 4-web samples 3-sphere

The deep-space frames used as background in the supplemental Blu-ray disc were provided by NASA and WMAP (Wilkinson Microwave Anisotropy Probe).

The supplemental HD videos contain the frames listed below

From Introduction on Blu-ray disc

From 1st HD video on Blu-ray disc

From 2nd HD video on supplemental Blu-ray disc

From 3rd HD video on supplemental Blu-ray disc

From 4th HD video on supplemental Blu-ray disc

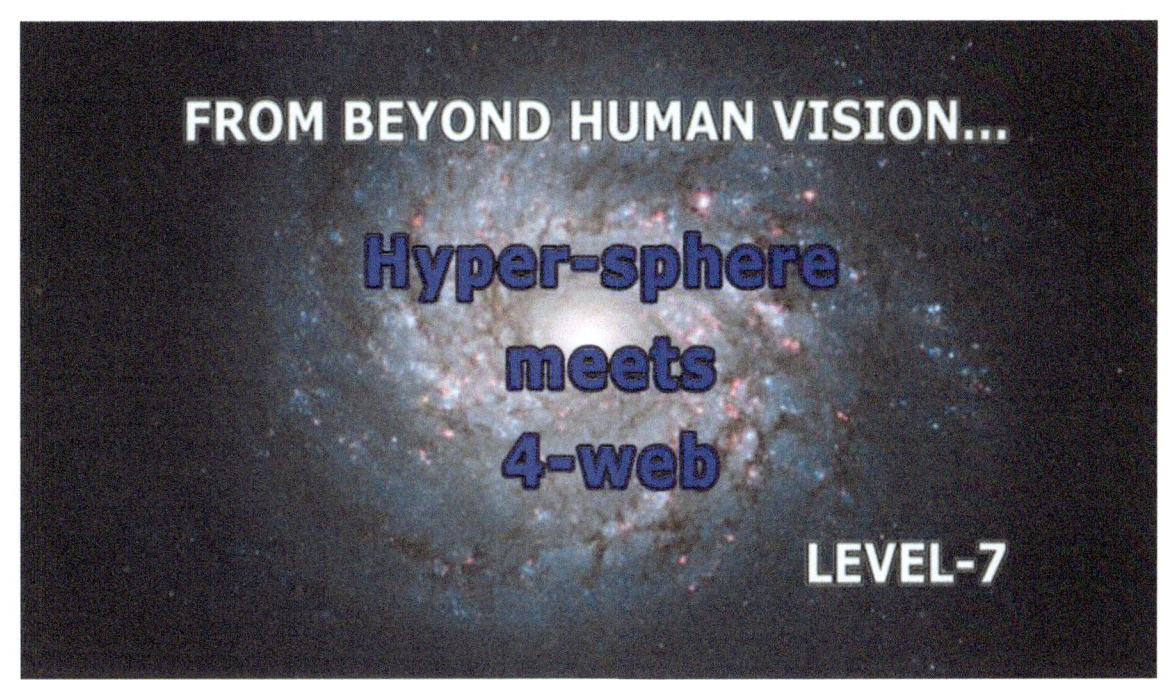

From 5th HD video on supplemental Blu-ray disc

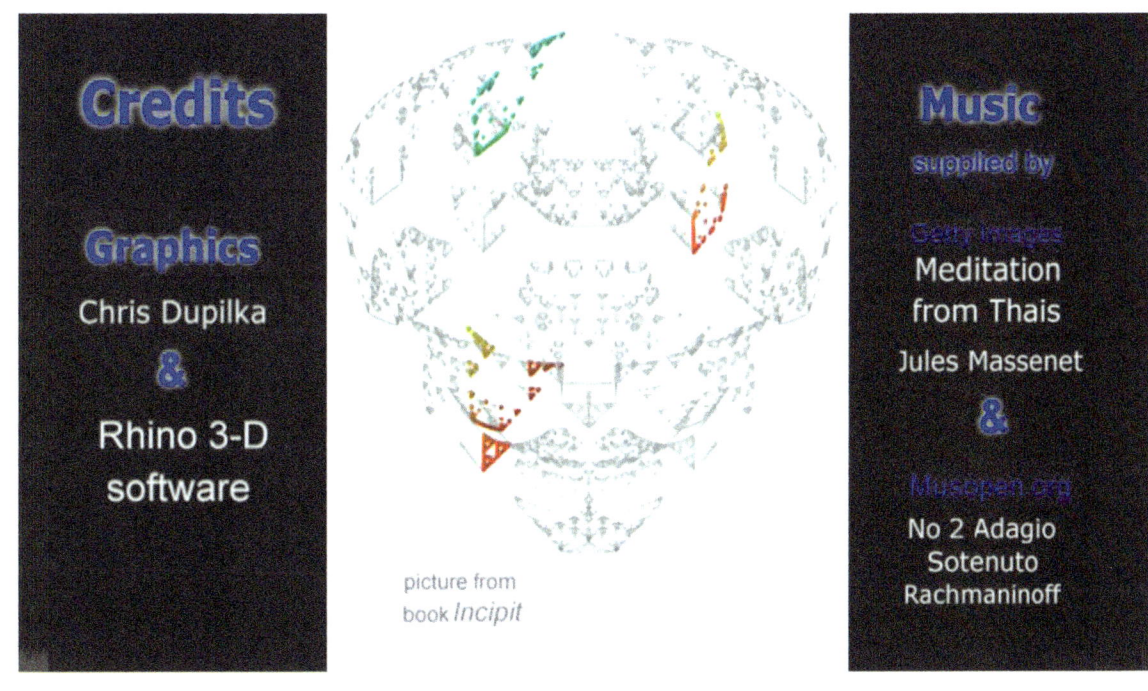

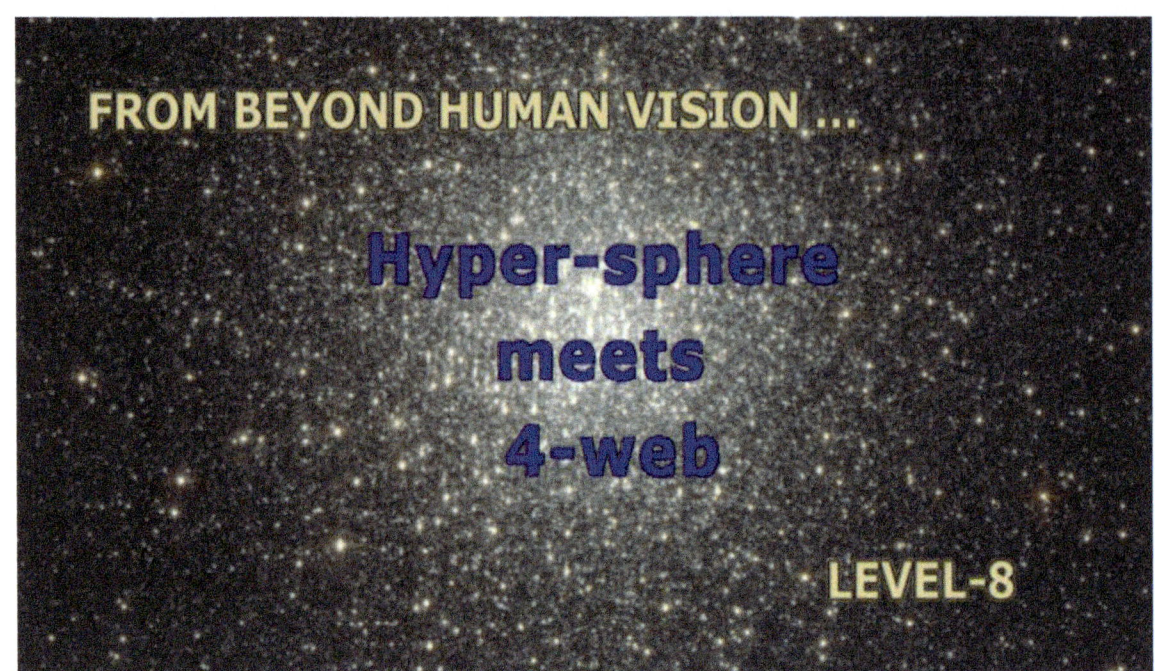

From 6th HD video on supplemental Blu-ray disc

CHAPTER 14

Appendix 3: Inside S^3 and Questions

This appendix provides an inside view of the 3-sphere and a few thoughts about the limits of physics.

§A19 INSIDE A 3-SPHERE

We understand our universe because we grow up in our universe, becoming ingrained with repetitive experiences. After a while, we view these experiences as *absolutes*. And then we deem any claim that deviates from our absolutes as *ridiculous*.

But history should teach us a lesson — think of the pre-Einstein days when we were ingrained with the idea that *time is absolute.*

In this section we shall describe visual experiences from inside a *rather small* 3-sphere. Keep in mind, however, that as the 3-sphere becomes increasingly larger *locally-repetitive experiences converge* to our everyday visual experiences.

To make the presentation interesting, we shall quote a few paragraphs on pages 32 and 33 in Volume 1 of William P. Thurston's 1997 book *Three-Dimensional Geometry and Topology*:[1]

> ... The easiest definition of S^3 is a unit sphere $x_1^2 + x_2^2 + x_3^2 + x_4^2 = 1$ in $\mathbb{R}^4$. Unfortunately, this formula does not immediately communicate a picture of S^3 to people who are not adept at visualizing four-dimensional space. But there is another way to imagine S^3, from the point of view of an inhabitant.
>
> To prepare the way, think first of what an inhabitant of S^2, the two-sphere, would see. By some mechanism, light rays are supposed to curve around to follow the surface. For instance, you can imagine that the "surface" is really a very thin layer of air between two large concentric glass spheres, which channel light by reflection in much the same way as fiber optics. (Unfortunately, the ecology of this model is not so clear. At best, there is just enough room for one to crawl around on one's stomach.)

---
[1] Thurston's book is published by Princeton University Press, Princeton, New Jersey.

Imagine creature A resting at the north pole, and another creature B creeping away. You can work out the visual images in terms of which geodesics (great circles) from the eyes of the A intersect B. As B creeps away, its image as seen by A at first grows smaller, although not quite as fast as it would in the plane. Once B reaches the equator, however, its image grows larger again with continued progress, until at the south pole, its image fills up the entire background of the field of vision of A in every direction.

Let us pause from Thurston's quote to illustrate his statements. We picture creature A with his eyes at the north pole looking at creature B.

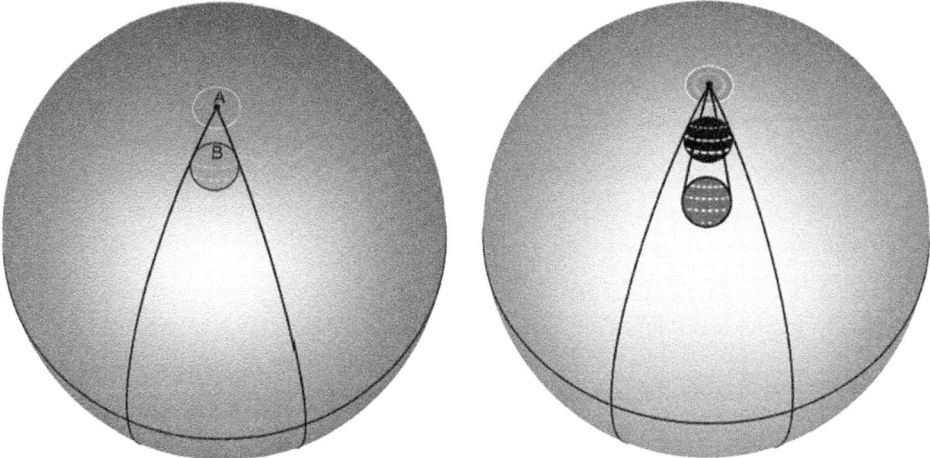

FIG. A3.1 From the viewpoint of creature A, the angle between light rays that travel along great circles to his eyes determines his view of *size*. As B (black spot) starts to move away, the size of B is determined by the two (partial) great circles that touch B. As B moves further away, creature A sees B (dark gray spot) as becoming smaller because the angle between the light rays (partial great circles) becomes smaller.

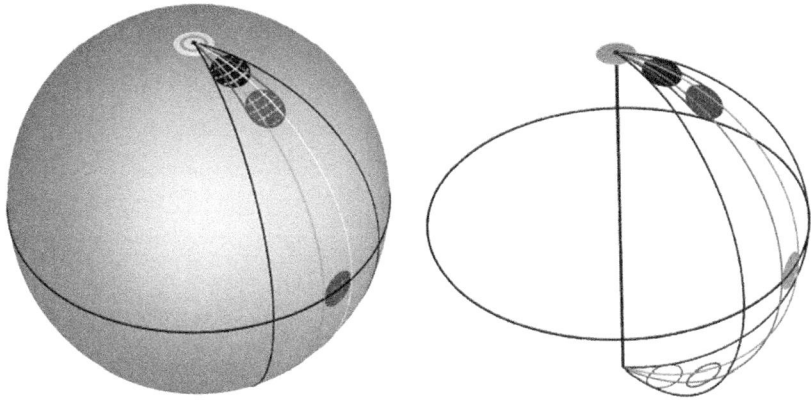

FIG. A3.2 On the left, creature B has now moved to the equator (light-gray spot), where A sees B as smaller (note the light-gray partial great circles). On the right, the surface of the sphere is removed, and we can see that as B gets close to the south pole, A sees B as the same size as when B was originally close to A.

For a view from the south pole, we include the following graphic.

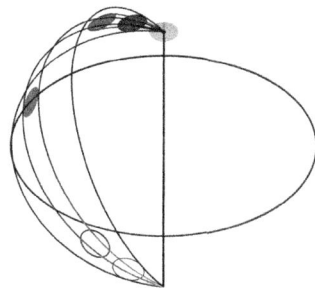

In the illustration above, note that the two "outside" half great-circles determine the angle at the eyes of creature A for both the near north pole and the near south pole positions of B. As B moves across the south pole, however, the light rays from B will come to the eyes of creature A from all directions — the image of B will, so to speak, fill the sky around creature A.

With the inside view of a 2-sphere in mind, let us return to quoting Thurston on the inside view of a 3-sphere (Figs. A3.1 and A3.2):

> The same phenomenon would take place in the three-sphere. Let's give ourselves more breathing room ..., and imagine that we are in a three-spherical world where a great circle is about two miles in circumference. There is no gravity, and we won't fuss about food, shelter, light or other minor details just now. We have little jets on our backs for flying around wherever we please. If I fly away from you, in any direction, my visual image to you shrinks in size at first fairly rapidly, but as I approach the

half-mile mark my visual size changes very slowly: it probably looks to you as though I have stopped making progress. After the half-mile mark, I gradually start to increase in visual size once more. As I approach your antipode, one mile from you, I start to grow rapidly again. When I am three feet from your antipode, the size of my visual image is exactly the same as if I were three feet from you. If I turn around and shout back, it will hurt your ears. We quickly learn that we can carry on a conversation with normal voices, for sound converges again at antipodal points just as light does.

Even though I have the same visual size to you when I hover three feet from your antipode as when I hover three feet from you, there is a difference in my visual image: you see further around to my sides. (There is also a difference in focal distance, but let's put that aside: imagine the light is very bright, so that your pupils are contracted and you don't notice this effect.) The difference becomes very dramatic if I now continue three feet further, so I cover the antipode of your eyes: you now see my image in every direction, and it is as if I were turned inside out onto the inside surface of a great hollow sphere totally surrounding you. You appear to me the same way, as the inside of a hollow sphere surrounding me.

In this description, we have left out an important part of the image. Light does not stop after traveling only a mile, it continues further. When I am a half mile from you, my image to you is as small as possible, but your lines of sight continue unimpeded completely around the three-sphere, to arrive back near where they started on yourself. In the background of everything else, you see an image of yourself, turned inside out on a great hollow sphere, with the back of your head in front of you.

There's another thing we left out: whenever I am at a distance other than one mile from you, you can actually see me in two opposite directions. For instance, when I was three feet from your antipode, had you turned around rather than me, you would have seen a perfectly normal image of me as if I were hovering three feet away and facing you, only slightly faded by the blue haze of the water vapor in the intervening air. You would also appear almost completely normal to me. But if we were to try to shake hands, they would pass through each other.

§A20 PHYSICS WITHOUT MEASUREMENTS?[2]

Underlying the very concept of *measurement* is the concept of *numbers*. And among all numbers, the number zero "0" is the most intriguing. The history of "zero" is documented in Robert Kaplan's book *The Nothing that Is, A Natural History of Zero* published by Oxford University Press in 1999.

Recall our school days when we were learning to place numbers one on top of the other, "lining up the digits" so to speak. Then drawing a line under the bottom number, we would subtract the bottom number from the top. And on occasion we would find that in the far right column we had a "3" on top and also a "3" on the bottom. At that point we needed a symbol for "3 − 3". Well try Roman numerals, "III − III". What Roman numeral symbol would you get?

The invention of zero underlies all of mathematics. Indeed, the *Null Axiom* of the ZFC axioms states: *There exists a set Φ with no elements*, i.e., a *set* (container) that contains nothing.[3]

So if Φ represents the number "0", then the set "$\{\Phi\}$" containing "Φ" is not empty. Indeed, "$\{\Phi\}$" contains one object and thereby represents the number "1". And so on. The *four sets* (four containers)

$$\Phi, \ \{\Phi\}, \ \{\Phi, \{\Phi\}\}, \ \{\Phi, \{\Phi\}, \{\Phi, \{\Phi\}\}\}$$

represent, respectively, the numbers 0, 1, 2, and 3.

In fact, all of the natural counting numbers can be so constructed. From these numbers, we recall from elementary school that we also study *negative numbers* as well as *fractions* and *decimals*.

The next step is to think of all of these numbers as *points on a line*. This is where we begin to connect *numbers* and *geometry*. Do you recall using a yard stick to measure length?. If so, then you used *numbers on a line*.

But once on a line, numbers are ordered geometrically, and we may talk about *convergence* — the idea of *a bunch of points somehow getting close to a given point*. For example, look at the list of fractions:

$$1, \ \frac{1}{2}, \ \frac{1}{3}, \ \frac{1}{4}, \ \frac{1}{5}, \ \frac{1}{6}, \ \frac{1}{7}, \ \frac{1}{8}, \ \frac{1}{9}, \ \frac{1}{10}, \ \ldots, \ \frac{1}{20}, \ \ldots, \ \frac{1}{40}, \ \ldots$$

Then look at their positions on a line ℓ:

---

[2] The science of physics is sometimes referred to as the *science of measurement* — the science of physics is based on measurements.

[3] For a concise presentation of the ZFC axioms see pages 362-368 of Mac Lane's book [22].

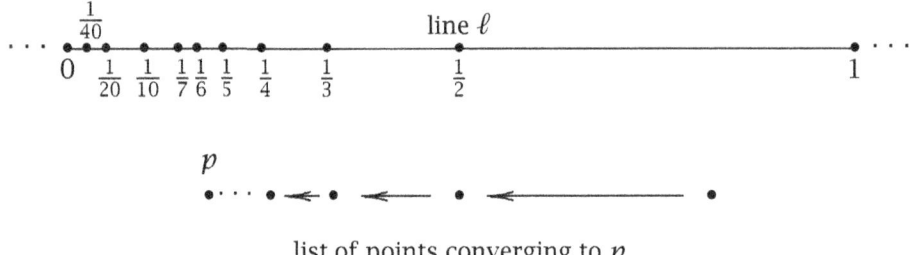

Fig. A3.3 An infinite list of fractions *converge* to "0". At bottom, an infinite list of points *converge* to a point *p*.

In Figure A3.3 we see an example of a *list of numbers converging to a number*. But thinking of numbers as points on a line, we can think about a *list of points converging to a point*. The idea of convergence seems to require measurements. When we travel in a car headed toward a bridge we often track the distance between "the point" where our car is located and "the point" where the bridge is located. And as we move closer and closer to the bridge, we experience the *distances (numbers) converging to zero*.

As mathematics evolved it became clear that there were many other structures, spheres for example, where we measured distances, and the distance between points on the surface of a 2-sphere can differ from the "straight-line" Euclidean distance. So convergence depended to some extent on how you would measure.

The term *metric space* became standard. A metric space is first of all a collection of points, say a plane, a sphere, or any collection of points that you can imagine. And second, a metric space also includes what is called a *metric* (meaning measure) which allows you to measure the distance between any two points.

With the idea of a metric space in hand, the idea of *convergence on a line* was generalized to *convergence in a metric space*. But the evolution of convergence was not over.

Within a branch of mathematics known as *general topology*, the idea of convergence includes convergence not only in all metric spaces, but also convergence in all *non-metric* spaces. Yes there are "spaces" where measurements between every pair of its points make no sense. The name for any such general structure is *topological space*. In fact, convergence is the essence of a topological space.

Formally speaking, a topological space is simply a collection of points where a certain *collection of its subsets* satisfy certain conditions.[4]

---

[4]For those with some background in mathematics, *convergence within topological spaces* in-

The point is simply that there are all kinds of examples of *spaces* that are *not metric spaces*, i.e., spaces where measurement of distances between points cannot be used to determine convergence. One classical example is the *extended long line*.[5]

So one naturally wonders whether at some future date, say when and if humans encounter a need to confront such *physical non-metric spaces*, will *physics extend its history as a science of physical laws within metric spaces to include physical laws within non-metric spaces.*[6]

---

volves generalizing convergence of points on a line. The classic reference is Chapter 2 *Moore-Smith Convergence* in John Kelley's 1955 book *General Topology* published by American Book Company in the University Series in Higher Mathematics.

[5]The extended long line is detailed as "Example 46" in Lynn Steen and Arthur Seebach's 1970 book *Counterexamples in Topology* published by Holt, Rinehart, and Winston, Inc. In addition, Steen and Seebach's book contains other examples of non-metric spaces. And in the vernacular of mathematics, most *function spaces* are not metric.

[6]For a gateway to historically relevant information concerning the emergence of various *kinds of spaces*, see Chapter 8 *Riemann's Legacy* of O'Shea's book, *The Poincaré Conjecture*, especially O'Shea's table on page 99.

CHAPTER 15

Appendix 4: Mathematics and Art

For background on "mathematics and art", let us quote Joseph Malkevitch :[1]

> ... there are, in fact, many arts (music, dance, painting, architecture, sculpture, etc.) and there is a surprisingly rich association between mathematics and each of the arts. ...
>
> One mathematical connection with art is that some individuals known as artists have needed to develop or use mathematical thinking to carry out their artistic vision. Among such artists were Luca Pacioli (c. 1145-1514), Leonardo da Vinci (1452-1519), Albrecht Dürer (1471-1528), and M.C. Escher (1898-1972). Another connection is that some mathematicians have become artists, often while pursuing their mathematics.
>
> Mathematicians commonly talk about beautiful theorems and beautiful proofs of theorems. They also often have emotional reactions to proofs or theorems. There are nifty proofs of "Dull" mathematical facts and "unsatisfying" proofs of "nifty" theorems. Artists and art critics also talk about beauty. Does art have to be beautiful? Francis Bacon's paintings may or may not be beautiful to everyone, yet there are few people who have no reaction to his work. Art is concerned with communication of emotions as well as beauty. Some people may see little emotional content in many of M.C. Escher's prints but it's hard not to be "impressed" by the patterns he created. Some find Escher's prints beautiful but with a different beauty from the great works of Rembrandt. Like art itself, the issues of beauty, communication, and emotions are complex subjects, but then so is mathematics.

History is replete with artists who have used mathematics. Arguably one of the greatest is Leonard da Vinci, e.g., see Bülent Atalay's *Math and the Mona Lisa* with subtitle "The Art and Science of Leonard da vinci" [1].

And the symbiosis continues. The evolution of *fractals* and *fractal art* — art that conveys a sequence of ever-smaller recursively-forming objects — provide such examples. In particular, consider M.C. Escher's July 1960 *Circle*

---

[1] See http://www.ams.org/samplings/feature-column/fcarc-art1 at the American Mathematical Society *Freature Column Archive*.

Limit V (Heaven and Hell), his 1959 *Circle Limit III*, and his *Smaller and Smaller* works. Describing his "Smaller and Smaller" piece, Escher states:[2]

> Repetition and multiplication — two simple words. However, the whole world of the senses would collapse into chaos without these two concepts. ... In a narrower sense they lead us to the subject of ... "regular division of the plane" ... In mathematical quarters, the regular division of the plane has been considered theoretically, since it forms part of crystallography ... If it counts as art, why has no artist — as far as I have been able to discover— ever engaged in it? ... I have never read anything about the subject by any artist, art critic, or art historian ... I must therefore attempt an explanation, and offer the following.
>
> *A plane, which should be considered limitless on all sides, can be filled with or divided into similar geometric figures that border each other on all sides without leaving any "empty spaces." This can be carried on to infinity according to a limited number of systems ....*
>
> The only way to escape this fragmentary character and to set an infinity in its entirety within a logical boundary line is to use the reverse of the approach in Smaller and Smaller.

In other words, the idea is that Escher's "Smaller and Smaller" is outlined with a basically-bounding circle, the larger figures appear at the bounding circle but become smaller and smaller as they move toward the center of the circle. In reverse, Escher's "Circle Limit" art displays the largest figures at the center as indicated below:[3]

And as the triangles approach the bounding circle they decrease in size. The illustration above exemplifies the "Circle Limit" view, an idea that mathemat-

---

[2] From the book *M.C. Escher, 29 Master Prints*, Harry N. Abraus Inc., New York, 1983.
[3] See http://mathworld.wolfram.com/PoincareHyperbolicDisk.html.

ically relates to the classical *Poincaré Disk*, which is well known and often pictured as an example of hyperbolic geometry.

Against this backdrop, one could rightly say that this book concerns "fractal art" because the construction of the art depends on the 2003 mathematical discovery of how to move the 4-web fractal from 4-space into 3-space with its structure preserved.

Is the meaning of "artist" being extended? One might wonder about the meaning of "artist" in the context of one whose brush strokes are guided by a fractal-based mathematical algorithm. Even more disconcerting is the fact that the dots on the canvas — thousands, if not millions, of dots in 3-dimensional space cannot be easily drawn by a human — the "paint color at various positions on canvas" must be selected by a computer. Nevertheless, from this author's view, given the advancement of computers and mathematics, the idea of looking into higher-dimensional space with the aid of new fractals seemed inevitable. And as far as this author knows, this book represents the first-ever presentation of applying a fractal-based computer algorithm to capture a partial image of a mathematical icon (the hyper-sphere) that lives only in 4-space.

Bibliography

[1] Atalay, Bülent, *Math and the Mona Lisa*, Smithsonian Books, Washington D.C., 2004.

[2] Asimov, Isaac, *Asimov's Guide to the Bible*, a 1293 page book published by Wings Books and distributed by Random House Value Publishing, Inc., 40 Engelhard Ave., Avenel, NJ, by arrangement with Double Day & Company, Inc., 1981.

[3] Barnett, Lincoln, *The Universe and Dr. Einstein*. Mentor Book published by The New American Library of World Literature, Inc. 501 Madison Avenue, New York 22, NY, Copyright 1948 by Harper & Brothers.

[4] Bing, R. H., The Collected Papers of R. H. Bing, American Mathematical Society, Providence, RI, 853–869.

[5] Blatner, David, *The Joy of π*. Walker and Company, NY, 1997.

[6] Buonaiuti, Ernesto, *Pellegrino di Roma: La generazione dell'esodo*, Laterza, Bari, 1964.

[7] Ciardi, John, in Dante, *The Paradiso*, Mentor, NY, 1970, notes to Canto 28, lines 21–36, p. 313.

[8] Crouch, Ralph, Walker, Elbert, *Introduction to Modern Algebra and Analysis*, Chapter 4, Holt, Rinehart and Winston, 1962.

[9] Daigneault, Aubert, and Sangalli, Arturo, Einstein's Static Universe: An Idea Whose Time Has Come Back? (A tribute to Irving Ezra Segal (1918-1998)), American Math. Society, Jan. 2001 *Notices*, page 9.

[10] Dante, *The Divine Comedy*, translated, with a commentary by Charles S. Singleton, Boilinger Series LXXX, Princeton University, Princeton, NJ, 1975.

[11] Dante, *Paradiso*, Harvard Univ., Cambridge, 1972, Canto 28, lines 58–60.

[12] Dugundji, James, *Topology*, Allyn and Bacon, Inc., Boston, 1967.

[13] Edgar, G.A., *Book Review*, Bulletin of the American Mathematical Society, Vol. 47, No.1, January. 2010, 163-170.

[14] Einstein, Albert, *Ideas and Opinions*, Dell Publishing Co., Inc., ISBN: 0-440-34150-7, Fifth Laurel printing — June 1981, Copyright © MCMLIV by Crown Publishers, Inc., Based on MEIN WELTBILD, edited by Carl Seelig, and other sources with New translations and revisions by Sonja Bargmann.

[15] Feynman, Richard, *Six Easy Pieces*, California Institute of Technology, 1995.

[16] Franks, M.R., *The Universe and Multiple Reality*, iUniverse, Inc., Lincoln, NE, 2003.

[17] Gardner, Martin, *Relativity for the Million*, 1962, Macmillon edition published November 1965.

[18] Kaplan, Robert, *THE NOTHING THAT IS, A Natural History of Zero*, Oxford University Press, 1999.

[19] Kelley, John, *General Topology*, American Book Company in the University Series in Higher Mathematics, 1955.

[20] Lipscomb, S. L., *Fractals and Universal Spaces in Dimension Theory*, Springer Monographs in Mathematics series, 2009.

[21] Lorentz, H.A., Einstein, A., Minkowski, H., and Weyl, H., with notes by Sommerfeld, A., Cosmological considerations on the general relativity theory", pp. 177-188 in the *The Principle of Relativity*, published unaltered and unabridged by Dover Publications, Inc., 1952.

[22] Mac Lane, Saunders, *Mathematics Form and Function*, Springer, 1986, pages 362-368.

[23] Nagata, J., *Modern Dimension Theory*, Bibliotheca Mathematica, Vol. 6, Interscience Publishers (John Wiley and Sons), NY, 1965.

[24] Nagata, J., *Modern General Topology*, Bibliotheca Mathematica, Vol. 7, North-Holland, Amsterdam, 1968.

[25] Nelson, Edward, *Mathematics and Faith*, http://www.math.princeton.edu /~nelson/papers.html.

[26] O'Shea, Donal, *The Poincaré Conjecture, In Search of the Shape of the Universe*, (pages 38, 39, and 206), Walker and Company, NY, 2007.

[27] Perry, J.C., and Lipscomb, S.L., The generalization of Sierpiński's triangle that lives in 4-space, Houston Journal of Mathematics, 2003, Volume 29, Number 3, 691–710.

[28] Peterson, Mark, *Dante and the 3-sphere*, American Journal of Physics, 47, 1979.

[29] Rhino, *Rhinoceros, NURBS modeling for Windows*, Version 3.0 Users Guide, Robert McNeel & Associates, 2002.

[30] Segal, Irving Ezra, *A variant of special relativity and long-distance astronomy*, Proc. Nat. Acad. Sci., Vol. 71, 1974, pages 765 to 768.

[31] Segal, Irving Ezra, *Mathematical Cosmology and Extragalactic Astronomy*, Academic Press, NY, 1976.

[32] Steen, Lynn A., and Seebach Jr., J. Authur, *Counterexamples in Topology*, Holt, Rinehart and Winston, Inc., 1970.

[33] Stillwell, John, *Yearning for the Impossible*, A. K. Peters, Ltd., 2006.

[34] Strobel, Lee, *The Case for a Creator*, Zondervan, 2004.

[35] *The Bible According to Einstein*, Jupiter Scientific Publishing Company, Columbia University, P.O. Box 250586, New York, NY.

[36] *The Holy Bible, Old and New Testaments, King James Version*, American bible Society, Instituted in the Year 1816, NY.

[37] Thurston, William P., *Three-Dimensional Geometry and Topology*, Volume 1, pages 32 and 33, Princeton University Press, Princeton, NJ, 1997.

[38] Weeks, Jeffrey R., *The Shape of Space*, Pure and Applied Mathematics, A series of Monographs and Textbooks, Marcel Dekker, Inc., N.Y., 1985.

[39] Wolf, Fred Allen, *Parallel Universes*, Simon & Schuster Paperbacks, 1988.

[40] *15th Edition of The New Encyclopædia Britannica*, 1974, page 485 of Volume 5 of the Macropædia.

Index

Symbols

0-cube, 53
1-cube, 53, 54
2-cube, 53, 54
3-cube, 53, 55
    face view, 54
    skew view, 54
    solid, 54
4-cube, 55
1-dim vision, 6, 122
2-dim vision, xv, 2, 6, 122
3-dim vision, 3
4-dimensional space, 53, 92, 112, 117
1-disc, 4, 122, 123
2-disc, 3, 122, 123
3-disc, 3, 16, 31, 123
4-simplex, 64
3s-sentence, 58
0-sphere, S^0, 1, 3, 9, 125
1-sphere, S^1, 1, 3, 8, 26, 41-43, 91, 125
2-sphere, S^2, 1, 3, 8, 28, 44-47, 49,
    91-93, 109-113, 119, 124, 150
3-sphere, S^3, 1, 8, 31, 32, 35, 41, 42, 69,
    91, 92, 121, 124-129, 163, 165
    bounded, 132
0-web, 55
    cell, 55
1-web, 55
    cell, 55
2-web, 55-62
    cell = triangle, 55
3-web, 55, 64
    cell = tetrahedron, 56, 64, 65
4-web, 50, 53, 55-62, 65
    camera position, 65-67
    cell, 55, 57
        edges of, 57
        equatorial triangle, 66
    equatorial triangular hole, 56
    equatorials, 56
    north pole, 56
    polar segment, 60
    solid hexahedron, 57
    south pole, 56
    density (segments), 62
    grid, 60-62, 64-66, 70, 73-76
    just touching, 57, 58
    moving into 3-space, 62-64
    subdivisions, 57, 59-62, 64, 70
        first, 56
        second, 59
        third, 59
        fourth, 60
        fifth, 70
        sixth, 61, 70
        seventh, 62, 70
        eighth, 62, 70
4-web fractal, 173

A

Afterglow Light Pattern, 140
Alexandria, 131
Angelic
    Choirs, 14
    sphere, 13-15
Angels, 13
antipodal points, 114, 115, 131
Aristotle, 1, 11
    Universe, 11, 12, 14, 16-18
Asimov, Isaac, 175
Atalay, Bülent, 171, 175
axiomatics, 131
axioms, 167

B

Bargmann, Sonja, 25
Barnett, Lincoln, 25, 133, 175
BBC, 86-88

Beatrice, 15-18, 21, 22
Bible, 70
Big Bang, 138-143
Bing, R. H., 33, 175
Blatner, David, 132, 175
Born, Max, 141
Buonaiuti, Ernesto, 175

C

Callahan, J. J., 129
Cartesian product, 35
    $R \times S^1$, 36, 37
    $R \times S^2$, 37-38
    $R \times S^3$, 38-39
characteristic equation, 106
Choice Axiom ("C" in ZFC), 131
Christian
    afterlife, 11
    life, 13
    soul, 12
    Trinity, 70
    wisdom, 22, 70
chronometric cosmology (CC), 39, 141-142
Ciardi, John, 21, 175
circle
    description, 131, 132
    finite length, 132
    perimeter, 132
consistency and spheres, 76-77
    lower-dimensions pattern, 76
convergence, converges, 167-169
converse art, 51
Cosmic Background Radiation (CBR), 140
Creation of Adam, 85-88
Crouch, Ralph, 175
cube, 44
curvature of space, 141
curved
    2-disc, 29
    equal-length segments, 27
    space, 141
    surface, 132

D

da Vinci, Leonardo, 171
Daigneault, Aubert, 35, 138-140, 175
Dante, ix, x, xvi, 1, 11-23, 70, 121, 128, 175
    journey, 12-14
    mirror, 16
    soul, 13
    3-sphere, 11-23
density of matter, 133
Descartes, René, 36
Dewhirst, David, 138
disc-shadows, 30-31
distortion, 42, 46
Doppler effect, 135, 137, 138
Doppler shift, 134-136
    changes in water pressure, 136
    colors of light, 137
    electromagnetic field, 137
    electromagnetic waves, light, 137
    field of air, 136, 137
    field of water, 136, 137
    frequency, 135
    frequency, pitch, 136
    galaxies moving, 137
    jiggled charge, 137
    jiggles the air, 136
    oscillating cork, 135
    oscillatory influence, 135, 136
    waves, 135
Doré, Gustave, 15-17, 23
Dugundji, James, 175
Dupilka, Chris, vi, x, 50, 61, 62, 64, 80
Dürer, Albrecht, 171

E

Edgar, G.A., iv, 57, 176
Eigenvalues, 95, 100, 106, 114
Eigenvectors, 95, 99-102, 104, 106, 107, 114
Einstein, Albert, vi, ix, x, xvi, 1, 22, 25-32, 35, 39, 132, 133, 138, 142, 176
Einstein's field equations, 134

Einstein's static universe, 35, 141, 144
Einstein's Universe (EU), xvi, 1, 35–39, 134, 140–141
Empyrean, 11, 13–17, 22
Equatorial circle, 45, 46
Escher, M.C., 171, 172
Euclid, 28, 131
Euclidean geometry, 26, 27, 31, 32, 131, 132

F

faithful dots image, 46
faithful representations, 42, 51
Feynman, Richard, 137, 176
Finlay-Freundlich, Erwin, 141
flex step, 121–123
    using a mirror, 123–124
flexed 1-disc, 121, 122
flexed 2-disc, 123
flexed 3-disc = hyper-hemisphere, 123, 124
four-dimensional space, 28
fourth dimension, 53, 54, 64, 65
fractal, 50, 64, 171, 173
fractal art, 171, 173
Franks, M.R., 144, 176
Freddy the Penguin, xv, xvii, 2, 9, 55, 62, 64
Friedmann, Alexander, 138
Fugier, Mary, x, 16
Füselli, Verlag Orell, 129

G

Gardner, Jimmy, x
Gardner, Martin, 176
general position, 74
geodesics. *See* great circles
gluing, xvi, 17, 121, 124
    1-discs, 4
    2-discs, 4
    hemispheres, 5, 18
    hyperhemispheres, 18
    locally glued area, 17–18
    semicircles, 4, 18
    two 3-discs, 5, 22
God, vii, x, 13, 22, 86, 87, 108, 143, 144
Golay, Pascal, x
graph paper, 41, 42
    cubical 8-cell grid, 45
    cubical grids, 45
    grid lines, 41, 45
    one-sphere, 41
    rectangular-cell grids, 43
    4-space, 42
    subdivision, 43, 44
    2-web, 43–44, 50
    2-web grid, 44
    3-web, 47–50
    3-web grid, 47, 49, 50
    4-web, 42, 53–64
Gray, John Earl, x
great 2-spheres, 91–119
great circles (geodesics), 91
group
    homomorphism, 115, 117
    kernel, 115, 117

H

Hambly, Jerry, x
Hawking, S.W., 143
Heaven, 12, 13, 22
heavenly spheres, 11, 12, 21
Hell, 12, 13, 22
hexahedron, 57
    just touching, 57–58
    solid, 57
hidden art, 85
Holy Scripture, 22, 70
Hoskin, Michael, 138
Hubble, Edwin, 134, 138, 141
human brain, 87–89
human face, 67
human skull, 67
hyperbolic geometry, 173

hypercube, 53-55
hyper-hemisphere = flexed 3-disc, 18, 123, 124
hyperplane slices, 8, 121, 126-128
hyper-space, 73, 74, 77, 79
hyper-sphere (3-sphere), ix, x, xv, 73-84, 108, 117, 119

I
Inferno, 12
invariant, 97
Ivanšić, Ivan, 144

J
Jerusalem, 12
"J" space, 144

K
Kaplan, Robert, 167
Kelley, John, 169, 176

L
Lehman, Larry, x
Lemaître, Georges, 138
light year, 134, 137
linear transformation, 92-93, 103
lines, 4, 26
    extended long, 169
    infinite length, 4
    line segment, 4
    straight, 168
Lipscomb, Darrin, 33
Lorentz, H. A., 38, 176

M
Mac Lane, Saunders, 167, 176
Mach, E., 133
Malkevitch, Joseph, 171
measurement, 167-169
meridian, 29
    circle, 45, 46
Meshberger, Frank Lynn, 86, 88, 89
metric space, 145, 168, 169
Michelangelo, 85, 86, 88
middle triangle, 43
Miller, Eugene LeRoy, v, x, 129
Milutinović, Uroš, 144
Minkowski, Hermann, 38, 142, 176

Minkowski space = $R \times R^3$, 142
Minkowski time = first factor "R" of $R \times R^3$, 142
mirror, 123-124
mirror image, 124
Moore-Smith convergence, 169
Mt. Wilson Observatory, 134
Musa, Mark, 15

N
Nagata, Jun-iti, 176
NASA, National Aeronautics and Space Administration, 139
natural numbers, 167
Nelson, Edward, 176
non-Euclidean geometry, 28, 31, 32, 131-132
non-faithful representations, 42
non-metric space, 168-169
Null Axiom, 167

O
O'Shea, Donal, 128, 129, 132, 169, 176
orthonormal basis, 101, 127
Osserman, R., 129

P
Pacioli, Luca, 171
Paradiso, 12, 13, 15, 21
parallel universes, 143-144
Parks, Allen Danforth, x
partial image of 3-sphere, 33, 67
    another view, 69
    chin, 68
    face, 67
    jaws/jaw lines, 67
    mouth, 68
    mustache (angel wings), 68
    nose, 68
perimeter, 132
Perry, James, v, vii, 50, 177
Peterson, Mark, 16, 20, 21, 129, 177
plane geometry, 26
Poincaré, ix, x, 108, 113, 119
Poincaré Conjecture, 128, 169
Poincaré Disk, 173

Poincaré Theorem, 142
predisposition, 86, 88
Primum Mobile, 12, 21
Projecting
    circle of constant latitude, 28
    great circle of constant longitude, 28
    S^1, 26-28
    S^2, 28-32
    stereographic projection, 26, 27
proof, 112, 123
Purgatorio, 13
Purgatory, Mount, 12, 13, 17

Q

quadratic form, 93-95, 100, 102
quantum physics, 143
quantum theory, 143

R

Real timeline, 36
red shift, 137-139, 141
relativity theory, general, 17, 133
relativity, theory of, 25, 143
Rembrandt, 171
Rhinoceros 3-D software, 9, 177
rigid discs, 31
Ritter, Albert, 12, 23
Robert McNeel & Associates, 9
rotation matrix, 95

S

saddle surface, 5, 132
Sangalli, Arturo, 35, 138-140, 175
Satan, 12
scaling, 67, 96, 104
science of measurement, 167
Seebach, Arthur, 169, 177
Seelig, Carl, 25
Segal, Irving Ezra, 35, 141, 142, 177

semicircles, 4
shadows, 26-31, 92, 93
shrink each cell by 1/2
    rectangular-cell, 42
    2-web cell, 43
    3-web cell, 47
    4-web cell, 56, 57
Sierpiński triangle, 44/cheese, 50
Sistine Chapel, 85, 86, 88
size of the universe, 133
slicing, xvi
    1-sphere, 7, 8, 18, 19
    2-sphere xiii, 8, 18, 91
    3-sphere, 9, 19-22, 121, 125-126
Slipher, Vespo, 138
Sommerfeld, A., 176
Speiser, Andreas, 129
spherical geometry, 25-28, 31
spherical universe, 133
Steen, Lynn, 169, 177
Stillwell, John, 12, 14, 15, 53, 64, 131, 177
straight-segment, 26
Strobel, Lee, 177
surfaces, 4, 131, 132, 163

T

tetrahedron = 3-web cell, 47
Thurston, William P., 163, 164, 177
topology
    general topology, 168
    topological images, 111, 119
    topological representation, 111, 113, 116, 119
    topological space, 168
translation, 96, 117
Trinity, 22, 70

U
universe
    density, 133
    spatially finite, 132, 133
    stationary, 141

V
Virgil, 22

W
Walker, Elbert, 175
Weeks, Jeffrey, 108, 119, 177
Weyl, H., 38, 176

Wikipedia, 23, 85
WMAP, Wilkinson Microwave Anisotropy Probe, 139, 140
Wolf, Fred Allen, 143, 177

Z
Zeleznock
    Janet, x
    Richard, x
Zermelo-Fraenkel (ZF) Axioms, 131
ZFC ("ZF" plus "C" axioms), 167

The manufacturer's authorised representative in the EU is Springer Nature Customer Service Centre GmbH, Europaplatz 3, 69115 Heidelberg, Germany. If you have any concerns regarding our products, please contact ProductSafety@springernature.com

Printed and bound by CPI Group (UK) Ltd, Croydon, CR0 4YY

26/03/2026

02078941-0020